Designing Production-Grade and Large-Scale IoT Solutions

A comprehensive and practical guide to implementing end-to-end IoT solutions

Mohamed Abdelaziz

BIRMINGHAM—MUMBAI

Designing Production-Grade and Large-Scale IoT Solutions

Group Product Manager: Rahul Nair
Publishing Product Manager: Rahul Nair
Senior Editor: Sangeeta Purkayastha
Content Development Editor: Rafiaa Khan
Technical Editor: Shruthi Shetty
Copy Editor: Safis Editing
Project Coordinator: Shagun Saini
Proofreader: Safis Editing
Indexer: Tejal Daruwale Soni
Production Designer: Jyoti Chauhan
Marketing Coordinator: Nimisha Dua and Sanjana Gupta

First published: May 2022

Production reference: 1170522

Published by Packt Publishing Ltd.
Livery Place
35 Livery Street
Birmingham
B3 2PB, UK.

ISBN 978-1-83882-925-4

www.packt.com

First and foremost, all thanks to God for helping me complete this book; then thanks to my wonderful wife, Shaimaa; my lovely daughter, Somia; my amazing sons, Omar and Zeyad; and the rest of my big family for supporting and encouraging me in writing this book, especially during the tough time we and the rest of the world have had due to the COVID-19 global pandemic.

– Mohamed Abdelaziz

Contributors

About the author

Mohamed Abdelaziz is a technology leader, IoT subject matter expert, cloud expert, and architect with more than 17 years of experience in IT and telecoms. He has designed and delivered many large-scale, production-grade, multi-million-dollar pieces of software and cloud-based solutions that cover both traditional IT and IoT solutions that are used by millions of users across the globe. Mohamed holds a degree in computer science and information systems, and he has gained multiple credentials in AWS (eight certificates) and Azure (five certificates – including the Azure IoT Developer certificate). Mohamed is an advocate of cloud computing, IoT, app modernization, containerization, and the architecture and design of large-scale distributed systems.

About the reviewers

Balaji is a Microsoft Azure IoT Specialty-certified IoT engineer and he holds a bachelor's in electronics and communications engineering. He has a significant amount of experience in the end-to-end processes of both the hardware and software sections of IoT. His area of expertise is in consumer electronics and automation. Quite recently, he expanded his horizons by stepping into Industry 4.0. He is currently exploring PLC, SCADA, and cybersecurity for OT in his free time. His areas of interest include but are not limited to low-power devices, data analytics, and ML on edge. When not tinkering with development boards, you can find him reading the New York Times bestseller of the month.

I would like to thank Packt Publishing for providing me with the opportunity to review this excellent book. I would also like to extend my gratitude to my mom and my friends who have stood beside me always.

Luca Palermo is an IoT practice lead and holds a BSc in computer science and an MSc in internetworking. Luca's career spans over 12 years in technology, including communication protocol design, solution architecture, and teaching at universities. Currently, he is employed with a system integrator, where he leads the designing and building of the largest Australian IoT platform. Luca is passionate about technology, in particular wireless networks, software design, and blockchain.

I would like to thank my wife, Giulia, for her continued support and encouragement with everything that I do, and my daughter, Sofia, for the unconditional love and joy she gives me every day.

Yatish Patil is an author and an Azure IoT analytics technology expert with a passion for building IoT solutions using Microsoft Azure. He works with enterprise customers, enabling them to identify and cultivate IoT analytics opportunities through technology innovation. His focus is on helping businesses accelerate their growth by using the cloud, mobility, and analytics with IoT, in which his career spans over 16 years.

He leads IoT practice and is actively involved in technology consulting and solutioning for customers, defining technology road maps, best practices, and processes for delivering on and achieving business objectives, consulting, and architectural aspects of end-to-end solutioning.

I'd like to thank my wife, Vasudha, and our children, Rudra and Rugved, for their daily support and patience. I'd also like to thank the Packt team for continuously giving me opportunities to review and write content on technology, which encourages me to continuously learn more and more and build a community through which we can learn as users. Also, to my friends, who encourage me in doing this stuff and appreciate and helpelp me from time to time on this exciting technology journey.

Table of Contents

2

The "I" in IoT – IoT Connectivity

3

The "T" in IoT – Devices and Edge

Section 2: The IoT Backend (aka the IoT Cloud)

4
Diving Deep into the IoT Backend (the IoT Cloud)

5
Exploring IoT Platforms

6
Understanding IoT Device Management

7
In the End, It Is All about Data, Isn't it?

Section 3: IoT Application Architecture Paradigms and IoT Operational Excellence

8
IoT Application Architecture Paradigms

9
Operational Excellence Pillars for Production-Grade IoT Solutions

Preface

Without any doubt, we are living in a true digital era, and the future will be even more digitalized. The keyword when we consider what it means to be truly digitalized is being connected. Many organizations and business enterprises are striving to make their products and solutions fully connected, with a view to gain more insights and operate their products both efficiently and remotely. The end goal is to increase their revenue, reduce operational costs, and increase customer satisfaction.

Internet of Things, or IoT for short, plays a vital role in connecting the physical products around us that we never imagined could be connected. Therefore, understanding IoT technologies and how they can solve different business problems, and having an understanding of how these technologies work is a must-have skill for any IT or software professional, both in our current era, and in the future as well.

People typically approach IoT technologies in one of following two ways - as a hobby, or as a profession. A hobbyist will typically use and play with IoT technologies to solve small problems they might have in their smart home, or in their office. They might purchase personal devices or microcontrollers such as Arduino or Raspberry Pi, along with some sensors. Then, using Wi-Fi connectivity and little or no code, they can program their devices to automate processes in their smart home and beyond.

Approaching IoT from a professional point of view is somewhat different, as IoT professionals (such as IoT solution architects, designers, or engineers) work in companies or enterprises that design and build large IoT solutions and products. Some examples of this could be smart cities, smart agriculture, connected vehicles, and so many other large IoT solutions that are available across thousands of different industries and business segments.

I wrote this book for IoT professionals who want to design and build large scale IoT solutions. I have worked in the IoT field for a long time now, so in this book, I'm sharing a wealth of practical experience in designing and building large IoT solutions.

There are three sections in this book. The first section covers the basics, or the foundations, of IoT, which is what I like to call the anatomy of IoT – for example, we will explore the meaning of the letters I and T in IoT.

The second section covers the IoT backend layer, or what I would call the IoT solution backbone. This is where we'll learn about the different software solutions and components that are used in that layer.

The final section covers the different IoT modern application architecture paradigms that are commonly used in almost all IoT solution layers. We'll also cover the IoT solution operational excellence pillars, which include IoT monitoring, security, resiliency, and availability.

In this book, I tried to cover a wide range of IoT topics, such as IoT connectivity, IoT device management, IoT analytics, IoT platforms, IoT applications, and many more. Our approach will be to first explain the concepts behind each topic, then show the different implementation options that you as an IoT solution architect or designer might come across during the IoT project designing and building phase. Finally, I have added my own design or architecture views, which is based on my practical experience and the technical pragmatic approach that is typically used in designing and building such large scale IoT solutions.

Who this book is for

The book targets E2E solution architects, systems and technical architects, and IOT developers who are looking to design, build, and operate E2E IoT applications and solutions. The book is not for beginners; however, some details and concepts will be provided in the book – where applicable – in certain sections/chapters.

What this book covers

Chapter 1, Introduction to the IoT – The Big Picture, gives an introduction to IoT and answers – in detail – three common questions about IoT (What? Why? How?).

Chapter 2, The "I" in IoT – IoT Connectivity, covers IoT connectivity in detail to help you understand and select the best IoT connectivity option for an IoT solution.

Chapter 3, The "T" in IoT – Devices and Edge, covers the hardware and software (Real-Time Operating System (RTOS)) aspects of IoT devices (microcontrollers) and IoT Edge devices.

Chapter 4, Diving Deep into the IoT Backend (The IoT Cloud), covers the IoT backend layer or the IoT cloud that is considered the backbone of any large-scale IoT solution. This chapter covers only the infrastructure part of the IoT backend layer by explaining the different options (that is, public versus private cloud, containers versus VMs, and so on) that can be used for hosting the IoT backend layer.

Chapter 5, *Exploring IoT Platforms*, covers the second part of the IoT backend layer, which is the software part. The chapter explains what is meant by an IoT platform, how to select the best IoT platform for an IoT solution, whether to build or buy an IoT platform, and the typical software components that should be provided by the IoT backend layer.

Chapter 6, *Understanding IoT Device Management*, covers the IoT device management capability in detail. The chapter explains the different IoT device management protocols commonly and widely used in the IoT domain, how to select an IoT device management solution, the typical features of IoT device management solutions, and how those features work and serve some business requirements of the IoT solution.

Chapter 7, *In the End, It Is All about Data, Isn't It?*, covers IoT data analytics in detail and how to design and build an IoT data analytics solution to gain the most desirable and valuable business insights out of the collected IoT data and other data from other data sources.

Chapter 8, *IoT Application Architecture Paradigms*, focuses on explaining new and modern application architecture paradigms such as cloud-native design concepts and the famous Twelve-Factor App methodology, microservice architecture, API gateways, service meshes, and other applications and architecture paradigms that can help you design and build (or customize) a robust IoT backend layer for a large-scale IoT solution.

Chapter 9, *Operational Excellence Pillars for Production-Grade IoT Solutions*, covers the pillars of IoT solution operational excellence. It covers pillars and aspects such as security, high availability, resiliency, monitoring, automation, and DevOps.

Chapter 10, *Wrapping Up and Final Thoughts*, summarizes what has been explained in the book by walking through an IoT industrial solution reference architecture and explaining its layers in detail. It also covers the future of IoT and the emerging cutting-edge technologies that will shape that future.

To get the most out of this book

Basic knowledge of cloud computing and distributed system design will help you to get the most out of this book.

Download the color images

We also provide a PDF file that has color images of the screenshots/diagrams used in this book. You can download it here: `https://static.packt-cdn.com/downloads/9781838829254_ColorImages.pdf`.

Conventions used

There are a number of text conventions used throughout this book.

`Code in text`: Indicates code words in the text, database table names, folder names, filenames, file extensions, pathnames, dummy URLs, user input, and Twitter handles. Here is an example: "To configure the host side of the network, you need the `tunctl` command from the User Mode Linux (UML) project."

A block of code is set as follows:

```
#include <stdio.h>
#include <stdlib.h>
int main (int argc, char *argv[])
{
    printf ("Hello, world!\n");
    return 0;
}
```

Any command-line input or output is written as follows:

```
$ sudo tunctl -u $(whoami) -t tap0
```

Bold: Indicates a new term, an important word, or words that you see on screen. For example, words in menus or dialog boxes appear in the text like this. Here is an example: "Click **Flash** from Etcher to write the image."

> **Tips or Important Notes**
> Appear like this.

Get in touch

Feedback from our readers is always welcome.

General feedback: If you have questions about any aspect of this book, mention the book title in the subject of your message and email us at `customercare@packtpub.com`.

Errata: Although we have taken every care to ensure the accuracy of our content, mistakes do happen. If you have found a mistake in this book, we would be grateful if you would report this to us. Please visit `www.packtpub.com/support/errata`, selecting your book, clicking on the Errata Submission Form link, and entering the details.

Piracy: If you come across any illegal copies of our works in any form on the internet, we would be grateful if you would provide us with the location address or website name. Please contact us at copyright@packt.com with a link to the material.

If you are interested in becoming an author: If there is a topic that you have expertise in and you are interested in either writing or contributing to a book, please visit authors.packtpub.com.

Share Your Thoughts

Once you've read *Designing Production-Grade and Large-Scale IoT Solutions*, we'd love to hear your thoughts! Scan the QR code below to go straight to the Amazon review page for this book and share your feedback.

https://packt.link/r/1838829253

Your review is important to us and the tech community and will help us make sure we're delivering excellent quality content.

Section 1: Anatomy of IoT

The objective of Section 1 is to help you deeply understand the fundamentals of IoT. In other words, deeply understand what the *I* and *T* mean in *IoT*.

This part of the book comprises the following chapters:

1

Introduction to the IoT – The Big Picture

In this chapter, we will start our journey of exploring the **Internet of Things**, or **IoT**, by introducing a number of topics, such as the purpose of any technology and the purpose of IoT technologies. We will then cover the three common questions (what, why, and how?) asked in relation to any technology, including IoT technology. We will cover what the IoT is, why the IoT matters, the impact of the IoT on different industries and segments, and finally, how IoT technology works. In other words, we will explain the IoT production-grade solution reference architecture and its solution building blocks.

The chapter will also cover some of the IoT solution design patterns commonly used in any IoT solution and will show a case study of a business problem and how IoT technologies can solve that business problem and deliver the desired business benefits and outcomes.

In this chapter, we will cover the following topics:

- The purpose of any technology (including IoT technology)
- IoT definition – the what
- IoT impact and benefits in different industries – the why
- IoT solution reference and architecture – the how

- IoT solution design patterns
- A case study
- The book's strategy

The purpose of any technology (including IoT technology)

Technology exists, and will keep evolving, to help achieve desired business outcomes. This is the ultimate fact. Business owners and investors always look for the following business outcomes when they run a company or invest in any business domain:

- Increased business sales, revenue, company value, the number of customers, and profits. In other words, an increase in the value of business growth with respect to the different business metrics we have just mentioned.

- Reduced business operational costs, which means the cost of running and operating the business must not exceed the value of the business profits by any means. Otherwise, the business is not considered profitable.

- The mitigation of different business risks such as new market disruptions and innovations, more competition, falling brand reputation (including an increase in the number of unhappy customers, extra product failures, negative social media feedback, and other factors affecting business brands), security risks, fraud risks, and operational risks (lack of automation, speed, and employee skills).

These are three common, well-known business outcomes every organization or business aims to achieve. There may be more, but this book does not cover economy or business theories.

Based on those aforementioned outcomes, it is clear now that no one will buy technology just to play with it; rather, they will buy technology that can help them achieve their desired business outcomes, or, in other words, they will buy technology that solves their business problems.

If you have been in business for the last few years, I am sure you've heard the term **Digital Transformation** a lot. Every organization nowadays is talking about its steps and strategies to transform its existing business into a digital-first business, especially after seeing the massive growth and business disruption of many large, digital, hyperscale companies such as Amazon, Facebook, and other tech companies that were born in the digital era, including Uber, Airbnb, and a host of other successful start-ups.

"Digital transformation marks a radical rethinking of how an organization uses technology, people, and processes to fundamentally change business performance", says George Westerman, MIT principal research scientist and the author of *Leading Digital: Turning Technology into Business Transformation.*

Many IT and tech companies offer business enterprises and organizations a complete, rich portfolio of services, solutions, and technologies that they provide to assess such enterprises in their digital transformation journey toward becoming digital-first.

Since this book is about technology (not *people* or *processes*), the question might be, how will technology help such a great digital transformation initiative? Or, in other words, what are the digital transformation technology enablers that each enterprise could use on such an exciting transformation journey? You might even ask why we are talking about business outcomes and digital transformation when this book is all about the IoT.

Regarding the last question, it is a valid one, and we will answer it soon. One thing to mention here is that we have decided to share with you the practical, or what is called applied, IoT experience that we have gained from being in the IoT field for many years. This experience stemmed from designing, delivering, and operating many large-scale and production-grade IoT solutions for many large enterprises around the globe:

Now, back to the question of what are the key technology enablers and solutions for enterprise digital transformation programs? And to answer this, let's look at the technologies and architecture paradigms that are considered key enablers for digital transformation:

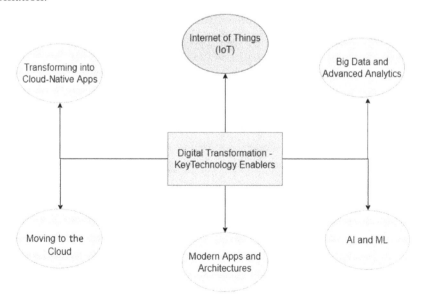

Figure 1.1 – Digital transformation – key technology enablers

Let's explain these key technology enablers in more detail.

Transforming into cloud-native applications and solutions

Transforming existing enterprise systems and IT workloads into new and modern cloud-native solutions is a key enabler in the digital transformation journey.

There are many different views on the definition of **cloud-native**, especially when it is compared with other terms and trends such as **cloud-readiness** and **cloud-enabled**. We will cover this in more detail later in the book. However, for now, we will stick to the official definition of cloud-native as per the **Cloud Native Computing Foundation (CNCF)**:

> *Cloud-native technologies empower organizations to build and run scalable applications in modern, dynamic environments such as public, private, and hybrid clouds. Containers, service meshes, microservices, immutable infrastructure, and declarative APIs exemplify this approach.*

The benefits of transforming organization workloads into cloud-native workloads are as follows:

- Releasing software fast so that it can hit the market quickly with new features
- Reducing costs, and leveraging the cloud and different pricing models, for example, pay as you go or paying for what you consume
- Increasing the scalability, reliability, and availability of enterprise applications and solutions
- Avoiding vendor lock-in – in other words, increasing application and solution portability

Moving to the cloud

There are many benefits of migrating and moving enterprise IT workloads and systems to the cloud (especially public clouds). We'll mention a few here:

- **Operational benefits, operational excellence, and operational efficiency**: Efficiency means you get more for less. In the public cloud model, the monitoring and support of application workloads become much more effective and manageable. Public cloud providers offer lots of automation tools, different support, and monitoring services to help enterprise operation teams.

- **Financial benefits and cost reduction**: The move from the **CapEx (Capital Expenditure)** model to **OpEx (Operating Expenses)**, or the pay-as-you-go model, helps you pay for what you use. With this model, by reducing the **total cost of ownership (TCO)**, you don't need to worry about data center facilities, ongoing hardware maintenance costs, and support to maintain the required reliability and availability of your IT and system workloads; the cloud provider will take care of such things.

- **Faster innovation or shorter time to hit the market**: Leveraging different cloud offerings, such as IaaS, PaaS, and/or SaaS, that enable the enterprise to try and build a **proof of concept (POC)** of any new ideas very quickly without worrying about buying hardware, software, platforms, and so on.

- **Manage business growth and regulations effectively and smoothly**: Most key public cloud providers offer a large set of regions and data centers across the globe. In a couple of minutes or less, you could deploy your workload in any region or geography around the globe if you have business needs in a particular region or have to observe local regulations or compliance requirements, such as data not leaving the country where the business is operating.

Big data and advanced analytics

Traditional analytics and business intelligence technologies and tools have long been used in enterprises and organizations of different sizes. However, the amount, velocity, and different varieties of data generated by business **IT (information technology)** and business **OT (operational technology)** systems change the way data and business analytic solutions work. Now, we are talking about **big data** technologies and **real-time or near-real-time data analytics solutions** versus historical data analytics.

AI and ML

Artificial Intelligence (AI) and **Machine Learning (ML)** play a vital role in the digital transformation journey as enterprises in different business segments use different AI and ML models and solutions, including predictive maintenance, product recommendations, personalized marketing and promotions, chatbots, drag discovery, fraud detection, radiology image recognition, video analytics, content commissioning, content creation, auto-subtitling, sentiment analysis, self-driving cars, and many other interesting business cases and solutions besides.

Modern applications and architectures

Creating a powerful customer digital experience requires the availability and utilization of a rich set of modern applications and architecture paradigms. For example, in modern apps, the de facto **architecture paradigms** commonly used nowadays are **microservice** and **two-tier architecture** versus the traditional three-tier architecture. When it comes to **compute**, modern solutions and architecture paradigms will include things such as **Serverless Event-Driven Compute** (for example, AWS Lambda, Azure, and Google Functions), **Containers** (for example, Docker), and **Container Orchestration** (for example, Kubernetes).

When it comes to **Middleware and Integration**, modern solutions and architecture paradigms will include things such as API Gateway, GraphQL, Service Mesh, Pub-Sub Architecture (queues and Service Bus), and Event Streaming (Kafka). When it comes to **Databases**, it is not just a question of traditional, well-known relational databases or **Relational Database Management Systems** (**RDBMSs**); other database types have emerged recently, including NoSQL, object storage, graph databases, and in-memory databases. When it comes to **Mobile Apps and Web Apps**, modern frameworks such as *React Native*, *Angular*, *Vue*, and many other frameworks and technologies besides will be used to build such amazing modern customer or consumer apps.

Internet of Things (IOT)

IoT is a key technology enabler for digital transformation. In fact, IoT is the most important technology enabler for the digital transformation process as it is also a key enabler for some other key technology enablers, such as big data and advanced analytics, AI, and machine learning. How? Simply put, such technology enablers depend heavily on collected data and insights from the different data sources of a running business. IoT is one such enterprise or business operational data source.

In the next few sections, we will discuss the different aspects of IoT, in other words, we will discuss the *what*, *why*, and *how* of IoT solutions and technologies.

IoT definition – the what

James Phillips, corporate vice president of Microsoft's Business Applications group, said the following;

It all starts with data; data is coming out of everything. If you can harness that data, make intelligence out of it, and use it to improve your business processes, you're in a position to transform your company and industry.

Have you heard about the **Observe, Orient, Decide, Act (OODA)** cycle that has been developed by a military strategist and United States Air Force Colonel, John Boyd? As the name suggests, it is a decision-making process based on a loop or series of steps or tasks. It all starts by observing or monitoring the target by collecting data from multiple sources, and then doing orientation by filtering, analyzing, and enriching the collected raw data. Next comes deciding, based on the insights collected and what actions need to be taken, and then finally executing such actions and assessing whether the decision taken was right or wrong. Such a loop continues endlessly.

Enterprises get insights from customer interactions, sales, and connected products to improve the quality of their products, find untouched revenue streams, increase customer satisfaction and retention, and improve the whole business process and operations.

As we can see, there is a strong relationship between IoT and data, so let's dive further into that relationship to understand IoT better.

The relationship between IoT and data

Imagine converting any physical thing around you into a smart connected thing and getting insights from that physical thing remotely. This would be awesome, wouldn't it?

How can I convert a physical thing into a smart thing, you may ask? The answer is by using IoT technologies and different IoT ecosystems, such as sensors/actuators, microcontrollers, connectivity, and IoT platforms.

Still not clear? Let's take an example. Suppose you run a waste disposal company. The trucks of the company currently travel once a week to different residential areas in the city. After a while, and based on some insights and data analytics, you realize that on some days, the trucks return almost empty as there is not much waste to be collected on those days. That is a business problem as you could save money on the fuel used and employee wages for those days, you could help to save the environment by reducing air pollution and toxic emissions, and so on.

To solve that problem, you need the waste bin to somehow talk to you or notify you whether it is full, halfway there, or empty. Interesting. So how can we do that? We could install in each waste bin a small low-power, constrained IoT device (or a microcontroller in short) equipped with an ultrasonic sensor (such as ultrasonic thru-beam sensors) to detect the level of waste inside the waste bin. You could have another sensor to detect how many times the bin lid has been opened. The microcontroller could have a radio communication module for short-range connectivity options such as Wi-Fi Bluetooth Low Energy or Zigbee, or long-range connectivity options such as cellular (mobile) network connectivity to provide internet or network connectivity to the microcontroller to send the sensor insights or data to the IoT Cloud backend.

Once you have the data in the IoT Cloud, then it becomes just another software or data analytics solution to get insights and act accordingly. Similarly, you could simply convert anything into a smart thing.

The concept of converting any physical thing into an internet-connected thing is really disruptive and has a big impact on the current internet infrastructure. Currently, there are billions of devices already connected to the internet, including laptops, mobiles, tablets, PCs, and other smart connected products, but with such a concept, we have to deal with massive growth in the number of devices connected to the internet network. In a single smart home, for example, you could have 100 devices or more connected to the internet.

The options in terms of dealing with such a massive number of internet-connected devices are either to extend the existing internet infrastructure or to have a dedicated IoT.

> **Definition of IoT as per Gartner**
> The IoT is a network of physical objects that contain embedded technology to communicate and sense or interact with their internal states or the external environment.

Now that we have an idea of what the IoT is, let's move on to its impact on different industries.

IoT impact and benefits in different industries – the why

Over the course of history, there have been different industrial revolutions that occurred in different eras. The first industrial revolution that occurred between the years 1760 and 1840 introduced **machine power** to replace hand power. At that time, this represented a significant achievement that disrupted and changed many industries, including mining, the iron industry, and agriculture.

Then, between the years 1871 and 1914, the second industrial revolution was enabled and powered by one of the greatest inventions and discoveries of humankind – *electricity*. Electricity fundamentally changed and disrupted different industries at that time.

In the late 20th century, the third industrial revolution was powered and enabled by the great invention of the *computer*. Computer systems and networking paved the way for the next and most recent industrial revolution, aka Industry 4.0.

The fourth industrial revolution (Industry 4.0) includes many technology enablers, such as **IoT, cloud computing, automation, AI, ML, analytics**, and **cybersecurity**, for disrupting a variety of industries and manufacturing processes.

IoT plays a critical role in shaping Industry 4.0 and impacts and disrupts any current industry and business. Let's learn how!

Manufacturing

With IoT technologies, different consumer products and services are not only classified as connected products, but they move to the next level of a business model where the following features can be easily introduced and bring great business value:

- **New billing and revenue models**: Think about smart bulbs, for example. What about buying smart bulbs on a monthly-based subscription instead of the current traditional model (that is, it is yours for just a one-off payment). In that new model, the consumer will not worry about ongoing maintenance issues with smart bulbs in their home as they will pay the monthly subscription to the smart bulb company, which will cover everything, including installation and ongoing post-installation maintenance. In other words, you rent the bulbs in your home instead of owning them. This new model enables and introduces IoT technologies, such as asset monitoring IoT solutions.

- **Increased quality of products and services**: A smart factory typically collects massive amounts of operational data that is generated by its connected devices, and also from the environment where those devices are deployed and operate. This helps the factory to undertake proper analytics on such massive amounts of data and introduce new features to the company's connected products or even predict failures in the connected products and act proactively in fixing those expected failures. This could be done by running predictive maintenance machine learning algorithms in the cloud based on the data collected by IoT sensors.

- **Increased business agility and resource optimization**: With IoT, companies or factory staff can run and maintain a company's connected products and services remotely using what is called a remote asset management solution. We saw during the COVID-19 pandemic in the year 2020 how companies that use IoT technologies in their products and services managed to survive and ensure business continuity during that difficult time. Also, with the concept of digital twins, workforce safety increased as staff no longer have to face some hazardous challenges during the manufacturing process.

Energy and utilities

Many challenges, such as connectivity issues, asset tracking and monitoring, safety, and security emanated from remote locations such as refineries and oil fields. Those challenges are addressed by leveraging IoT technologies such as IoT device management, IoT connectivity, edge computing, ML, and Analytics @EDGE.

Also, on the consumer side of things, IoT helps a lot in saving energy. Think about smart lights that detect a person's movement in a room and switch on only if there are people in the room. They can also detect brightness levels – whether it is daytime or during the night, and act accordingly. Other examples that employ the same idea are air conditioning systems and heaters. These kinds of solutions save lots of energy resources and reduce consumers' bills as well.

In the utilities sector, things such as smart meters are the first requirement for many utility companies.

Transportation

The IoT fundamentally impacts and disrupts the transportation industry. IoT technologies play a critical role in the following services and features in that industry:

- Connected vehicles.
- Autonomous vehicles and self-driving.
- Fleet management.
- Increased safety and security.
- Increased efficiency.
- Onboard entertainment.
- And many other interesting features and services besides. Recently, as a response to the COVID-19 pandemic, we have started to see trains and buses being equipped with multiple additional sensors to help train and bus companies enforce social distancing and other health-related rules that were put in place to help fight the pandemic.

Health care

The IoT fundamentally impacts and disrupts this industry. IoT technologies play a critical role in the following services and features of the industry:

- Wearable patient devices to remotely monitor blood pressure, heart rate, and many other biometrics. This helps in tracking the health status of old and lonely people to act immediately in the case of an emergency.

- Doctors can now diagnose their patients remotely as they get all the data that they need to collect from their patients through attached IoT devices or sensors. With the introduction of 5G (a super low-latency connectivity option), it will become possible to perform remote surgery, which will help save lots of lives in remote areas.

- Different hospital equipment can be tracked and monitored by IoT using IoT asset tracking solutions.

Retail

Have you heard about the interesting "Amazon Go" stores that have no staff working in them? Consumers just pop into the store and buy whatever they want and when they leave the store, payment is taken automatically? Interesting, isn't it? IoT and other modern technologies are behind those types of smart stores.

In retail, the IoT supports many use cases, such as the following:

- Real-time inventory management
- Personalization
- Safety and security, that is, video analytics
- Cost savings and operational efficiencies
- And so many other interesting features and services

Sports and leisure centers

The IoT supports many use cases in the sport and leisure center domain, such as the following:

- Real-time tracking
- Improved safety with video analytics
- Interactive fan engagement

Agriculture

IoT technologies introduce many benefits to the agriculture and farming sector, such as the following:

- Livestock and crop tracking.

- Improved crop quality and increased volumes.

- Increased farming process efficiency. Think about smart irrigation and how much water the farmer could save if they manage to detect soil moisture and see whether crops require watering. Also, in a smart irrigation solution, you could get input about the weather status and whether it will rain, so you could save even more water on this basis.

One of the interesting use cases in that sector was a case of a farmer complaining about an increase in the newborn cattle death rate (this is the business problem) that occurred because the farmer did not know exactly when cattle would give birth. During labor, human intervention may be required to help female cattle should a problem arise.

So, how could IoT solve that business problem? With a very tiny microcontroller equipped with a vibration sensor and a cellular communication module attached to the tail of the pregnant cattle, at a specific vibration rate (this specific rate indicates that the cattle is giving birth), a video call will be triggered with the farmer and the latter can watch the whole process from their home and intervene if needed. Through this solution, the volume of newly born cattle has increased significantly.

Supply chain

IoT technologies introduce many benefits to the supply chain sector, such as the following:

- Detecting counterfeit goods

- Improved product quality, that is, better control of products' expiration dates

- Real-time or near-real-time inventory management

Cities

Resource scarcity, or a lack of resources, occurs when demand for a natural resource exceeds supply. Therefore, in a modern economy, there is always a need for new and sustainable financial and service models.

Smart cities are a business initiative to offer a sustainability model and the best quality of life standards with fewer resources.

IoT technologies help a lot in achieving the vision of smart cities and completely disrupt lots of traditional citizen and council services. Here, we'll mention a few:

- Smart parking
- Smart street lighting
- Smart waste management
- Smart energy management
- Smart buildings

And so many more features and services.

Home automation – smart homes

With IoT technologies, the concept of a true and complete smart home is possible. Many products and services in homes are powered by IoT technologies. Here, we'll mention a few:

- Smart refrigerators, microwaves, ovens, coffee machines, TVs, and so on
- Smart lighting
- Security and safety
- Smart windows, doors, curtains, and so on
- Smart thermostats
- Smart gardens

And many other features and services besides.

Without any doubt, and as has already been stated, the IoT is everywhere and disrupts many different business sectors.

> **Important Note**
>
> What is interesting here is the fact that the IoT solution paradigm is very simple and very common. In any sector, you will have things or IoT devices equipped with sensors to sense the physical world and with communication modules to provide connectivity to send the collected generated data from those IoT devices to the IoT backend cloud for further analysis.

To conclude this section, the most important thing is understanding business problems, how IoT can solve these issues, and what benefits businesses will get from using IoT technologies. Let's now move on to understand more about IoT solutions and technologies and see how it all works together.

IoT solution reference and architecture – the how

When an architect, designer, or developer starts thinking about solution architecture for a business problem, they usually start thinking about a standard reference model or reference architecture that has been used and tested, with proof of success, by other experts in solving the same business problem they have in hand.

In traditional IT systems, you go with well-known architecture paradigms such as **three-tier architecture**, **Service-Oriented Architecture (SOA)**, **two-tier architecture**, and many more modern paradigms, such as **microservice-based architecture** and **serverless architecture**. Even at the software code level, there are software design patterns.

IoT solutions are not unlike traditional IT solutions in the sense of the need to have a standard solution reference architecture or solution building blocks for any IoT solutions. However, IoT solutions are different in the following ways:

- IoT solutions, by default, are End-to-End (E2E) solutions, from devices/sensors to web and mobile apps for end user use.

- IoT solutions contain so many architecture paradigms that in the IoT solution application layer, you could leverage one of the architecture paradigms we mentioned before, for example, three-tier architecture or serverless, while in the IoT solution analytics layer, you have different architecture paradigms, such as Lambda and Kapp architecture, with other IoT solution layers. We will explain those types of architecture paradigms later in this book.

- The footprint of skills and technologies required for IoT solutions is big compared to traditional IT solutions.

In *Figure 1.2*, we have tried to capture, to some extent, the standard or commonly agreed upon IoT reference architecture, or IoT solution building blocks, that can be used to address IoT solutions for different business problems in different business domains:

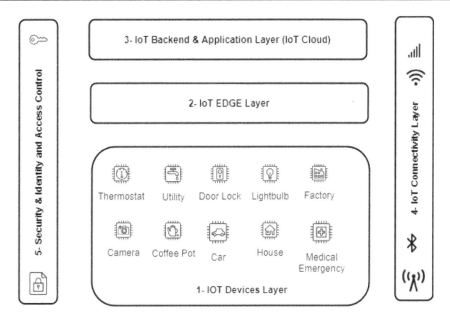

Figure 1.2 – IoT solution reference architecture

Let's examine each layer of the IoT solution reference architecture in some detail.

IoT devices layer

This layer of the solution focuses on the IoT devices and their ecosystems. IoT devices can be classified into two categories:

- **IoT endpoint devices**: This kind of device is usually cheap, low-power, or battery-based, such as constrained microcontroller devices that have sensors attached to them to sense objects in the physical world, or a wireless communication module to support short-range connectivity options (Wi-Fi, Zigbee, Bluetooth, and so on) to connect that device to an IoT edge or gateway device or to the IoT backend cloud directly using long-range connectivity options such as cellular, Ethernet, or satellite connectivity options. They also have a **real-time operating system** (**RTOS**) and different software stacks, **Software Development Kits** (**SDKs**), and embedded systems.

- **IoT gateway devices**: This kind of device is more powerful or bigger in terms of resources (compute, networking, and storage) compared to resource-constrained and limited IoT endpoint devices. Such types of devices usually run gateway services and have features such as the following:

  a) They act as a communication hub/router, that is, helping IoT endpoint device-to-device communication or IoT endpoint device-to-IoT backend cloud communication.

  b) Data caching, cleansing, buffering, and aggregation locally at the edge.

  c) IoT endpoint devices go through the gateway to get internet access and don't directly connect to the internet, so such devices manage IoT devices' security.

  d) They play a role in the edge computing paradigm as sometimes it is used for local data processing and analytics.

  e) The IoT gateway plays a critical role in supporting legacy and protocol conversion. Old IoT devices typically run old IoT connectivity protocols. The IoT gateway can convert those old protocols to modern and supported IoT protocols that run in the IoT edge layer or the IoT Cloud.

We will cover this layer in detail later in the book.

IoT edge layer

The edge or fog computing paradigm, in short, is running data center workloads very close to the end user or devices. You can move a small data center workload or an entire workload to the edge; it all depends on the edge location facility size and the data center's supported capabilities.

Let's look at the three main drivers behind the need for the edge computing paradigm.

Latency

We can't beat the speed of light, right? This is physics. In other words, whatever the quality and strength of a fiber optic cable used for data transmission in a packet data network, the transmission of the data will not be done in zero or fewer milliseconds.

Think about an IoT device running in a car park in the Singapore region, with the IoT analytics and applications running in the IoT backend cloud in the USA region. We should expect – and we can't do anything about it – in the region of a 250 ms latency factor added to the overall application request-response latency. So, a request raised from the IoT device in the car park might take roughly 1 or 2 seconds in total (250 ms latency + application processing time + database processing time, and so on) to get a response, assuming the workload running in the IoT Cloud is well designed and implemented and will not add any further unnecessary latency to the response time.

In some applications, getting an answer or response from the IoT backend cloud in 1 or 2 seconds could be fine, but in other real-time or near-real-time applications, that time could be a big problem. Regarding the example we just discussed, imagine there is a fire in the car park and the drivers need to get out or escape as soon as possible. The car park gate is closed and waiting for a response from the IoT backend cloud to open. Getting the response in 1 or 2 seconds in this case (note: we are not discussing here a no-response scenario as that is a completely different case altogether) might be too late to save people's lives.

In network topology, data traffic goes through multiple nodes or hops till it reaches its destination. Besides the light speed latency that we discussed above, you could also have a number of network issues, such as the cable being torn or some congestion in one of the nodes in the network topology. All such factors could even make the situation worse in terms of latency or getting a response quickly from the IoT backend cloud.

Edge computing solves these issues and challenges by running the IoT Cloud workload required (for example, Apps, Analytics, ML, and Control) locally, that is, as close as possible to IoT devices.

Cost savings

Data is important and, in fact, the goal of any IoT solution is to acquire data, gain some insights, and finally, act upon those insights.

IoT devices generate, or can generate, massive amounts of data. You can have data generated in a very small time resolution, for example, one second or less. Domains such as analytics and ML usually require such big data to give proper analytic outputs or excellent ML model accuracy results, but processing and storing such an amount of data comes at a cost.

Edge computing solves that challenge by processing all (or part of) such IoT-generated data at the edge and storing only the relevant and required data needed for further analysis and applications in the IoT backend cloud.

In the edge cloud, you can do data aggregation first before sending the data or batch of data to the IoT backend cloud. You can run advanced near-real-time or streaming analytics on the edge and might be storing just the results of such real-time analytics in the IoT backend cloud for other historical analytics solutions (for example, trends). For example, running analytics on the last 10 or fewer minutes of usage of smart parking could be done easily on the edge with data stored and processed on the edge for such a time resolution, that is, 10 minutes. After that, data could be deleted or aggregated into a higher time resolution (for example, 1 hour or so) and that aggregated data is then sent to the IoT Cloud for historical analytics.

Edge computing solves the preceding issues and challenges by reducing the cost of storing and processing such massive amounts of generated IoT data in the IoT Cloud.

Data locality and privacy

Due to certain regulations or data privacy compliance requirements, you might have to store sensitive data such as personal data generated from IoT devices within the country or region where those IoT devices are deployed and operating. A problem may arise if you have a kind of centralized IoT backend cloud in one or more regions that differs from the highly regulated regions the IoT devices are deployed and operated in.

Edge computing solves that issue by storing and processing such sensitive data locally and might anonymize such data and push it later to the central IoT Cloud in an anonymized or masked form.

We will cover this layer in further detail later in the book.

IoT backend and application layer (IoT Cloud)

This layer is very important in any IoT solution as it covers so many solutions and applications. Let's look at these in detail.

Provisioning layer

Provisioning, in general, means setting up and configuring backend systems with all the required information and configurations the solution's upstream and downstream systems require to operate as expected.

In IoT solutions, IoT device provisioning can be done in many backend systems depending on the final IoT solution architecture, but here are common systems required in IoT device provisioning:

- **A thing or IoT device provisioning**: This system or platform is responsible for storing IoT device metadata in the IoT Cloud database – usually, it is called an IoT device registry solution. Metadata such as the device ID, device description, and so much other metadata that you could store about the IoT device will help you in the solution later; for example, storing the device's location (which floor of the building, parking lot, and so on the IoT device is installed on) might help in an end user's journey when it comes to searching for a smart parking solution and suchlike.

  Also, IoT device identity details such as device credentials and/or X.509 certificates can be securely stored and provisioned in that layer.

- **Connectivity provisioning**: The IoT device might have one or more communication modules, for example, one communication module for Zigbee connectivity and another one for cellular (mobile) connectivity.

  In the case of cellular connectivity, for example, you must configure or provision a **Subscriber Identification Module** (**SIM**) with the mobile network operator, otherwise such connectivity will not work.

Ingestion layer

This layer is the front-door layer of backend IoT Cloud solutions and services. In other words, this is the first layer that receives data from the IoT endpoint devices or IoT edge and ingests such data into the proper IoT Cloud storage layer. In that layer, you can have the following components:

- **MQTT Message Broker**: We will cover the **Message Queuing Telemetry Transport** (**MQTT**) protocol in greater detail later in this book, but for now, MQTT is a lightweight publish-subscribe (or Pub-Sub) network protocol that supports message transportation between devices, so one device publishes the message to a specific topic and the other device(s) or applications from the other end subscribe to that topic to get the published message. MQTT is considered one of the best IoT application communication protocols for a wide variety of reasons, which will be discussed later. But for now, the MQTT lightness feature is the most obvious reason to prefer MQTT over HTTP as IoT endpoint devices are usually constrained in terms of computing resources, so running heavy protocols such as HTTP might be a problem or not supported at all by the IoT endpoint device operating system.

A complete IoT solution should have a scalable, reliable, resilient, and secure MQTT message broker, or we can call it an MQTT server since the IoT endpoint, IoT edge devices, and IoT applications usually act as MQTT clients.

- **Streaming Processing Engine**: Data, in the case of powerful IoT devices, could come from the IoT devices directly in the form of data streams over the HTTP(S) protocol or, typically in large-scale and production-grade IoT solution architecture, you have in the front an MQTT message broker and that message broker sends or forwards such incoming IoT data (coming through MQTT) to the streaming processing engine for further processing if required.

A complete IoT solution should have – if needed – a scalable, reliable, resilient, and secure streaming processing engine such as Kafka to support IoT data streams.

IoT rule engine

This component is critical and crucial in any IoT solution. Why? Because this component is the glue between IoT devices on the ground and the IoT backend cloud. In other words, the component is responsible for directing incoming data from the IoT devices to their destination in the IoT backend cloud. There are so many destinations, such as the database (SQL, NoSQL, and so on), message queue/bus, data lake, or object storage such as Amazon S3, Hadoop, and a streaming engine for further processing or real-time analytics use cases.

Storage layer

In this layer of an IoT solution, there are so many different storage options, as we will discuss later in the book, but the most common one used in large-scale and production-grade IoT solutions is object storage solutions such as the famous Amazon S3 object storage service.

The concept of the data lake – usually built on top of object storage solutions such as Amazon S3 – is the recommended IoT design pattern where all data in whatever format coming from IoT devices will be ingested and stored durably and securely in that data lake storage for subsequent processing.

Further down the line of IoT solutions, you could have another process or system read such raw IoT data from the data lake for further processing and storage. For example, you could perform some data cleansing, preparation, and processing, and store the data in another data store such as a SQL database (for reporting) or a NoSQL database (for a real-time dashboard application).

Analytics and machine learning layer

This layer of the IoT solution covers all systems and components used in building a big data and analytics standard pipeline (collect->process->store->analyze->visualize).

It also contains systems and components used in building an ML pipeline (Data Extraction -> Model Training -> Model Evaluation -> Model Deployment).

This layer is so important because, as explained earlier, IoT is all about getting the data for analytics and insights.

We will cover this layer in further detail later in the book.

IoT applications layer

In this layer of the solution, there are many components. Let's briefly discuss them.

Compute services

You will write application code and you will need to deploy or run that code, so you will require compute services to host the application code. These are the typical options:

- **Bare-metal or physical host**: This option is so expensive and not often used in the era of public cloud and managed hosting services. It could be used only if you have legacy application code that has special hardware specifications or special software restrictions, for example, licenses.

- **Virtual Machines**: This option is the most used compute option whether applications are deployed in a traditional data center, a private cloud, or a public cloud.

  Virtualization technologies have been a game-changer in the computing service domain in recent decades and are still valuable options in terms of modern application deployment.

- **Containers**: Container technologies are the latest compute services that offer the greatest benefits. We'll cover them later in the book.

  Container technologies help a lot in achieving the desired benefits of new application architecture paradigms such as microservice architecture. Microservice architecture concepts were not new, but with container technologies, they become brighter and make much more sense.

Container technology introduces a need for a container orchestration platform to manage container deployment on a large scale. Kubernetes is an open source container orchestration engine designed to deploy and manage containerized services on a large scale.

There are other container orchestration platforms available on the market offered as commercial solutions, such as AWS ECS – Elastic Container Services, or open source, such as Apache Mesos or Docker Swarm. However, Kubernetes has proven to be the best and is the market-leading container orchestration platform with huge technical community support.

- **Serverless**: Serverless means there is a server, but the *less* part of serverless means you do not manage that server at all.

  In the serverless paradigm, the application developer will focus on the code only, be it Java, Python, C#, and so on. Then, when it comes to deploying that code onto a server-side platform, the developer simply uploads the code artifacts to the serverless provider (whether it is a public cloud provider or a private cloud provider). The serverless provider behind the scenes will deploy that code to a server managed by that provider.

  Serverless technologies, also called **Function as a Service (FaaS)**, typically run and manage containers at scale behind the scenes to deploy and run the user's uploaded code.

  Serverless usually follows an event-driven architecture paradigm. To execute the uploaded code, an event must be triggered to notify the serverless service to execute your code and return its result, or it can be triggered in a scheduled manner.

  Examples of serverless offerings are AWS Lambda, Azure Functions, and Google Cloud Functions.

As part of the IoT solution design phase, you have to choose the compute service you require for your solution. You might prefer containers over virtual machines or vice versa, you might choose both for different applications' requirements, or you may prefer serverless for a quick start and to avoid server maintenance, and so on.

Database services

In modern applications, no one database engine or type fits all data and applications' purposes. Currently, there are lots of database engines for different uses and purposes, such as the following:

- **Relational databases or SQL databases**: This is the oldest and most well-known, with RDBMSs such as Oracle DB, Microsoft SQL Server, My SQL, MariaDB, PostgreSQL, and many more. Such kinds of database engines are usually used in traditional applications, ERP, and e-commerce solutions.

- **Non-relational (or NoSQL) databases**: As the name suggests, this other type of database engine emerged to solve some of the limitations associated with relational database engines, including scalability and availability.

There are many forms of NoSQL databases, including the following:

a. Key-value databases are commonly used in online gaming apps and high-traffic web apps.

b. Document databases are commonly used in content management systems and user profiles.

c. Graph databases are commonly used in social networking, recommendation, and fraud detection apps.

d. In-memory databases are commonly used in caching, session management, and gaming.

e. Time series databases are commonly used in IoT and industrial telemetry apps.

f. Ledger databases are commonly used in supply chain, registration, and banking transactions.

We will discuss these databases in more detail later and shed light on which one to choose during the design phase of an IoT solution. The IoT solution architect or designer should evaluate the different database options for the IoT solution based on the requirements at hand.

Middleware and integration services

This layer of the solution provides integration or middleware systems to connect IoT solution systems and integrate third-party and external systems as well.

There are many integration systems available, as per the following:

- **Legacy middleware platforms**: Such platforms were introduced earlier to support the integration of **Service Oriented Architecture** (**SOA**)-based applications. Such platforms host many integration services or features such as service bus, orchestration (the **Business Process Execution Language** (**BPEL**) engine), and many more integration features are included in the middleware platform, making it heavy or monolithic.

   You might need one or two features out of all the included or built-in features of a single middleware platform. This makes decision makers think a lot before choosing between such kinds of platforms or the modern option, that is, an API gateway.

   It is worth mentioning that companies behind such legacy middleware platforms have taken serious steps to modernize them to fit with new microservice and cloud-native architecture paradigms.

- **API Gateways**: With the new microservice-based architecture and API-first architecture paradigms, the need for a lightweight middleware-like component increased to avoid microservices' direct communication and to provide gateway features in terms of security, routing, and tracing. The API gateway was the component introduced into modern application architectures to offer those features. An API gateway can also be classified as a strong or a smart gateway.

- **Event Bus / Message Queues**: Message queues introduce significant benefits in terms of decoupling and scaling application components.

- **Stream Processing Engine**: The Pub-Sub (or publish-subscribe) pattern is one of the microservice-based application communications patterns. The other microservice communication pattern is the Sync pattern, where a microservice calls each other microservice's exposed APIs directly over the HTTP(s) protocol.

   Pub-sub fits very well with microservice-based applications as microservices are usually developed, deployed, and scaled in standalone mode. Each microservice can trigger or broadcast events related to its operations and, based on the events triggered, the other microservices interested in such events can subscribe and listen to those events and act accordingly; for example, an order-created event triggered by a cart microservice. Then, other microservices, such as payment, inventory, and credit checkers that are interested in, and listening to, those kinds of events (that is, order-created events) can then start their functionality on the created order.

Apache Kafka is the most well-known streaming processing engine used on the market.

- **Service Mesh**: A new technology introduced to help service-to-service communications, the service mesh concept came into the picture to work side by side with API gateways in a very good and complete integration architecture pattern where an API gateway is used for external integration, while a service mesh is used for internal service-to-service communication. We will explain microservices, service meshes, and API gateways later in the book.

IoT applications and visualization

This layer hosts IoT applications, visualizations, and dashboards. In other words, this layer is the layer responsible for building what the end user will see and interact with, so it is an important layer in the whole IoT solution.

An E2E IoT solution is complex and requires many systems and solutions in order to be delivered. But if you think about IoT end users or consumers who will use the IoT solution in the end, you'll realize that those end users will interact mainly with the IoT applications of the solution in the form of a mobile or web app. Hence, the excellent customer experience that every enterprise or organization looks for will be driven mainly by that IoT application layer's **Key Performance Indicators** (**KPIs**).

To give an example, in your E2E IoT solution, you might have some problems in the connectivity layer, so, for the end user in that case, you could offer what is called a digital twin or device shadow to let the end user keep interacting with the IoT solution as normal, as if the IoT device is still connected. And when the device reconnects, it will read the instructions sent while it was offline from the device shadow service and apply what is needed. Or, let's say you have a problem in the device layer; for example, devices are not reachable at all and no telemetry is received from devices in the configured time.

In this instance, the entire IoT solution shouldn't go into stop or failure mode; you could still offer IoT application layer components to the end user, such as mobile and web apps, until you fix the device issue, but in that case, they will be in read-only mode, that is, the dashboards and reporting could still be offered to the end user. Yes, it will show out-of-date data, but that's better than shutting down the whole solution and losing end user engagement. The user might just want to check some historical reports or something.

In this layer, there are so many solutions and systems to be considered in building and delivering IoT solution applications. Let's briefly go through such components.

Dashboards and visualizations

There are many solutions when it comes to building IoT solution dashboards and visualization. Here, we'll mention a few:

- **Ready-made IoT dashboard solutions**: In this kind of application, IoT developers will be given a platform that has many built-in controls and widgets and they just need to configure or customize those controls and widgets with the IoT data source. This is a drag-and-drop kind of development.

  There are commercial and non-commercial IoT platforms that provide such features as ThingWorx (commercial) and ThingsBoard (open source) and many others.

- **Low-code solutions**: Another option could be using generic low-code solutions to build the IoT solution dashboard and visualizations. Platforms include Microsoft PowerApps, Mendix, Appian, and many others besides

  Usually, the two options above, that is, ready-made dashboards or an application built by a low-code platform, are used to quickly start the development cycle of IoT applications without the need for highly-skilled backend or full stack software developers to build the IoT applications from scratch.

- **Do It Yourself (DIY) Apps**: In this option, the company or the software vendor's developers will build the IoT solution dashboards, visualization, and apps from scratch.

  Here, the IoT developers will use different frameworks and solutions to build IoT apps, such as the following:

  a) Web and mobile frameworks – frontend.

  b) Technologies and frameworks such as JavaScript, React, Angular, Vue, React Native, Ionic, Flutter, and so many others.

  c) Microservices and backend APIs.

  d) eTechnologies and frameworks such as Spring Boot with Spring Cloud, Flask, NodeJS, .NET, and so many other frameworks and technologies that are available on the market to build microservice-based applications.

  e) GraphQL (read-only) works side by side with API gateways (read/write). It makes it easy for applications to get the data they need efficiently without chattiness between a client (for example, the browser) and backend server or backend API.

Now that we've discussed IoT applications and visualizations, let's move on to exploring IoT device management applications.

IoT device management applications

Managing an IoT device in terms of device provisioning, configuring, maintaining, authenticating, and monitoring is one of the mandatory IoT solution requirements. It is rare to find an IoT solution without a device management component in its architecture and ecosystems.

There are two options for IoT device management solutions:

- **The buy option**: There are lots of IoT device management solutions available on the market that cover most of the device management requirements and features.

- **The build option**: You can build an IoT device management solution in-house or with a software partner.

Without connectivity, there's no IoT solution. Connectivity is everywhere in an IoT solution and it acts as a glue between all IoT solution layers. Let's look at this in the next section.

IoT connectivity layer

Solution architects and designers should cover different connectivity options required in the IoT solution and how to connect IoT endpoint devices to edge devices and edge devices to the IoT backend cloud, which should cover which wireless (or wired) technology and communication protocols are to be used.

This layer will be detailed later in the book.

Security and identity and access control

Like connectivity, security and access control is a must-have requirement in any IoT solution. Different IoT solution components must incorporate security requirements into their design and delivery from day 1 or day zero. An IoT security breach is massive and dangerous in its impacts. Think about the hacking of connected cars and what could happen if a hacker has full control of a car while you are driving. What about switching off smart city streetlights or switching off electric grids and so on? It is serious, isn't it? We are now talking about systems that affect people's lives directly, not just traditional websites and online services.

Solution architects, designers, and developers should include all the required security and access controls in all IoT solution components. Topics such as authentication, authorization, malware protection, auditing, access control policies, and data protection in transit and at rest should be fully covered in the IoT solution.

Those are the five layers or the solution building blocks of any IoT solution. There are some other additional technological frameworks and tools used across all those layers, but they are mainly part of, or driven by, the delivery process used in the delivery of the IoT solution. For example, if the organization you are working in follows DevOps or DevSecOps practices, then developers and/or DevOps engineers will use things such as the following.

Infrastructure as code

Infrastructure as code means treating the infrastructure of the IoT solution the same way you treat the solution's source code. In other words, the infrastructure code or scripts are maintained in version control the same as the solution's source code, which will give us lots of benefits, such as faster time to production/faster time to market, improved consistency, less configuration drift, and modern app deployment methods. And finally, it improves the automation of the entire solution deployment.

There are many tools used for infrastructure as code. The most famous infrastructure or cloud-agnostic and open source tool is Hashicorp Terraform.

CI/CD (Continuous Integration / Continuous Delivery) pipelines

A solution or project's CI/CD pipeline tools usually include the following:

- **Source code repository or version control**: This is where the project or the solution's code artifacts will be stored and managed. Infrastructure code or scripts can also be hosted in this repository. There are so many source code repositories available on the market. The most common and famous repositories are those built in a Git repository, such as GitHub or GitLab.

- **Build and test tool**: When it comes to the build stage of the CI/CD pipeline, we have so many options and tools available to build the solution code and other solution artifacts, such as Maven, Ant, Docker, Gradle, Packer, and many more tools available on the market.

  Usually, unit test scripts and integration scripts run as part of the building stage.

- **Continuous Integration tool**: This tool is the brain or the orchestrator of the whole CI/CD pipeline. It triggers the pipeline once developers commit code or based on a specific schedule in the Git repository, and then it executes the build and testing scripts. It then deploys the project binary and artifacts generated from the build stage to testing, staging, and finally to production environments if configured to do so, that is, without manual reviews. There are many tools, such as Jenkins, CircleCI, Bamboo, TeamCity, and others available on the market to do that job.

- **Artifact repository**: This is a tool to store and manage project or solution artifact deliverables. Usually, they are packaged in binary format. Such packages are basically the output of the build stage of the CI/CD pipeline. Examples include Docker images and the binary of the software. There are many tools on the market, such as Artifactory, Nexus, and many others in that domain.

In the next section, we will discuss IoT solution design patterns.

IoT solution design patterns

In software coding practices, you should have heard about software design patterns, in particular, those patterns introduced by the famous **Gang of Four** (**GOF**) book *Design Patterns: Elements of Reusable Object-Oriented Software*.

The idea of design patterns, in general in any domain, is to provide a solution that has been tested and proven by different experts to solve specific repeated problems or challenges in the subject domain. For example, in the software domain, you will see a list of design patterns that solve some common coding challenges, such as the Factory pattern, the Singleton pattern, and so many other patterns. In the enterprise integration domain, you will see patterns such as Publish-Subscribe, Event-Driven, Content-Based Router and Message Filter. In the cloud, you will see patterns such as Ambassador, Anti-Corruption Layer, Cache-Aside, and Responsibility Segregation (**Command Query Responsibility Segregation** (**CQRS**)).

In the IoT domain, we do have some design patterns and design principles that are commonly used in different IoT solutions. Let's look at some of these.

Telemetry

This is the most famous and common pattern used in IoT solutions. IoT devices sense the physical world and send related data to the IoT Cloud for further processing. Data sent from devices is called data telematics or telemetry.

To send the data to the IoT Cloud, you need connectivity and a communication protocol. Different protocols may be employed, including HTTP(S), WebSocket, and MQTT.

Since we are talking about *telemetry*, **MQTT** is the best and most common protocol used in sending IoT device telematics to the IoT Cloud or, to be more specific, to the IoT Cloud Message Broker or Edge Device gateway running the MQTT server.

This telemetry pattern helps many companies to build and offer IoT SaaS solutions or ready-made solutions, where enterprises buy ready-made IoT solutions to have their products connected very quickly and start to gain useful insights from them. The idea is simple: connect the device to an IoT platform and start sending telematics and the IoT platform will provide dashboard, monitoring, alerting, reporting, and analytics features out of the box based on the telematics data collected. Microsoft Azure IoT Central is one of those IoT application platforms that offer such ready-made IoT solutions with no great development skills required.

Command

This pattern is also very common in IoT solutions and usually comes side by side with the telemetry pattern. The command or action pattern means applying or running different commands to a remote IoT device – commands such as *reboot*, *reset*, *switch on*, and *switch off* device actions, and even upgrading the device firmware, is also considered a command pattern.

It is usually best practice to have control and management of the remote IoT device for monitoring, diagnostics, and security.

Outbound connectivity only

IoT devices are not like traditional web or proxy server devices where you can open some inbound ports such as 80 or 443 for incoming traffic from the internet or from a private network. Opening such inbound ports for traditional powerful devices could be acceptable, as usually there are firewalls, **Distributed Denial-of-Service** (**DDoS**) protections, and **Web Application Firewall** (**WAF**) solutions in place to protect such proxy web servers or the whole server farm from such attacks.

In the IoT, it is slightly different. IoT endpoint devices usually, or in most cases, do not run an IP stack, meaning there's no IP assigned to devices. They do not run an IP stack as they are low-power, constrained, and low-value devices. Even in the case of powerful IoT endpoint devices running an IP stack, it is recommended to not open its inbound ports to mitigate security attacks. Security breaches in IoT solutions are more dangerous than a breach in e-commerce applications or any other IT solutions as the impact of an IoT security breach will be severe on systems and humans as well.

So, the most common IoT device communication mode is the outbound communication mode (aka device to cloud), where the device will make an outbound call to the external network instead of receiving inbound calls (aka cloud to device) from external networks.

You might wonder how the command IoT design pattern we mentioned above works then? As explained before, a command means you send instructions to an IoT device to do some actions. Hence, this means inbound traffic from your application/system toward an IoT device. So how come we say IoT device connectivity is usually, or recommended to be, outbound?

This is a very good question and the answer is simple. The IoT device makes an outbound call to ask about any actions or instructions targeted for it that need to be executed. As we will learn through the book, IoT devices might be in a deep sleep mode most of the time and when they wake up, they communicate with the IoT Cloud or the IoT Edge; *Hey, I am up now. Any jobs for me?* Then, the IoT device will subscribe to a specific topic for such jobs and retrieve instructions and execute them.

Refer to the next pattern as this is related to that point as well.

Device twin or device shadow

For sure, connectivity will be dropped or disconnected at some point in time, and you should always consider that fact when designing an IoT solution. IoT devices might be deployed in very far away locations where there is no network coverage at all, where there is weak network coverage, or where intermittent connectivity occurs due to one of many reasons, such as electronic or radio interference from other surrounding devices or even the IoT device being in deep sleep mode to save the battery, meaning it is not connected all the time.

The solution to that intermittent connectivity challenge is to use an intermediate system or buffer to hold and manage the communication between the IoT devices from one end and the IoT application from the other end. In other words, IoT applications should not talk directly to IoT devices and vice versa. Rather, IoT applications should talk to the intermediate system and instruct it on what needs to be done on the IoT devices, and then that intermediate system will talk to the IoT devices if the IoT devices are on – the best-case scenario. Otherwise, if they are off, then that intermediate system will hold the message until the IoT devices come back to life, wake up, or connect again. Then, they will get the message buffered in the intermediate system when they were off and act accordingly.

Each IoT platform implements the above concept of the intermediate system. AWS calls it the *Device Shadow service*, Microsoft Azure calls it *Device* Twin, and so on.

Device bootstrapping or device provisioning

When you buy smart consumer devices, there is usually a common step to configure some device settings before the device works, settings such as the device access point or device connectivity, which Wi-Fi the device should use to connect to the internet in the case of Wi-Fi-enabled devices, which **Access Point Name** (**APN**) should be used in the case of cellular LTE connectivity, and which Bluetooth peer should be paired with. This process is called device setup, device provisioning, or device bootstrapping.

In **industrial IoT** (**IIoT**), provisioning or bootstrapping IoT devices is different and challengeable. In IIoT, we are talking about a massive number of IoT devices, so how can we have a device bootstrap on such a scale?

To give an example, think about a connected product company that produces thousands or even millions of such connected product devices. They sell devices across the globe. Now, you need a way to streamline the bootstrapping process for all of those IoT devices. If you choose cellular connectivity in your connected products, then how will you bootstrap or provision the SIM inside the device in different countries with different mobile network operators? It is a bit tricky and not a straightforward process.

Usually, two topics emerge in the provisioning and bootstrapping process.

Device identity

As a manufacturer or owner of IoT devices, how can I make sure that only my IoT devices – not any other IoT or non-IoT devices – talk to my IoT Cloud or my customer's IoT Cloud? There's a need to have some sort of attestation or credentials that are stored securely in IoT devices in order to connect to the IoT Cloud. The device will send such credentials to the IoT Cloud during the connectivity handshake, which can then be validated and authenticated by the IoT Cloud.

IoT device credentials could be in the following forms:

- A password or token
- An X.509 certificate
- A **Trusted Platform Module** (**TPM**)

There are pros and cons of each option, which we will cover in detail later in the book. But for now, and regarding bootstrapping, the question is how those credentials are stored and used by the IoT device.

To answer that question, a myriad of options are available, including the following:

- During the IoT device manufacturing process

- Over the air, that is, pushed to devices after manufacturing

- **Just in Time** (**JIT**) provisioning, that is, when the device connects for the first time to the IoT Cloud

- Dependent on a third-party solution, that is, a telecommunication network backbone

We will cover these options later in the book.

Device home endpoints

Where should the IoT device connect to send the data telematics? Will such an endpoint be statically configured in the device or will it be pushed to the device over the air? Is there a global point an IoT device can and must connect to when it connects for the first time, and then that global point will or might direct the IoT device to the endpoint the IoT device should connect to, for example, a specific region or country's IoT hub or IoT message broker?

All those questions should be answered and clarified as part of the device bootstrapping process.

Now that we have covered IoT solution design patterns in detail, it's time to discuss some use cases to broaden our horizons.

A case study

Despite this book being generically about designing and building an IoT solution, we will look at a case study to be used in the book's chapters where applicable.

A smart *city* initiative is not one IoT project; it is, in fact, so many different IoT projects and it usually has the following cases or modules to be covered:

- Waste collection management

- Street lighting management

- Digital building management

- Energy management

- Green space management

- Forest fire detection

- Digital citizenship services

- Smart parking

In this book, we've chosen the smart parking module of the smart city initiative to be used as a reference or case study:

Figure 1.3 – Parking lots (source: https://wordpress.org/openverse/image/d621a8cf-e786-4877-adea-bc5eb0ce7f87)

In the upcoming sections, we will go through the requirements or the business problems at hand and the design thought process of an IoT solution to address those business requirements and deliver the desired business benefits.

Parking space issues

Usually, in large, bustling cities around the world, finding a parking space (aka slot) is challenging and not an easy task at all. People sometimes avoid outings on account of parking issues. Governments and city council administrations usually look for solutions to address parking availability challenges. There are so many factors behind parking space issues, such as *ineffective land planning, insufficient parking spaces, traffic* congestion, *toxic emissions*, and others.

Putting ineffective land planning issues aside, as we can't help with that issue, the congested traffic and toxic emissions issues occur essentially because car drivers lack a real-time parking lot availability solution to enable them to efficiently and quickly find available parking lots in the city.

The solution to the parking issue

City administrators started looking into smart parking solution initiatives to gain the following business benefits and solve parking space issues:

- Reduce toxic emissions and help save the environment.

- Reduce traffic congestion.

- Reduce costs with less fuel consumption.

- Reduce the amount of time drivers spend searching for an available parking lot.

- Increase revenue – with smart parking, parking lots will be occupied efficiently by different drivers throughout the day, and a new billing model will be introduced, such as paying *for exactly how long you stay*.

- With smart parking analytics and ML technologies, city council administrators can undertake proper land planning to increase the number of parking spaces across the city.

Business requirement analysis

As stated earlier, understanding business problems and pain points in full is an important step toward solving such business problems. Sometimes the solution might be a non-technical solution and might not depend on technology at all. The solution might be just enhancing existing business processes or people skills.

In that phase of the project, that is, business requirement clarification, you, as a business or system analyst or solution architect, should put any questions you have to the business stakeholders as answering such questions will have an impact on the solution selected at the end. For example, *Is the parking space indoors or outdoors?* You might ask, will this matter? And the answer is yes, it will matter, as, based on the answer, you will define which devices, sensors, and connectivity options could be used. For example, if it is indoor parking, then you could have a source of electrical power available in the parking lot or nearby so you might not worry much about low-power, very constrained IoT devices and so on. As regards the sensors, you might not worry much about the external environmental and non-environmental impact on the functionality of the sensors, such as dust and lighting levels. For connectivity, you might choose Wi-Fi and fixed broadband instead of cellular (mobile) connectivity and so on.

The business and system requirement identification process is not an easy task; it would need a book to explain that topic at length. However, here are the general steps used in that process:

- What are the smart parking solution use cases? What are the business drivers for a smart parking solution or what are the goals of the business related to the smart parking solution?
- Who are the stakeholders of the smart parking solution?
- What are the business's smart parking and technical KPIs?
- What are the smart parking solution's **Non-Functional Requirements** (**NFRs**), such as security, availability, scalability, disaster recovery, and resiliency?

Let's dive a bit deeper here and discuss the role of various stakeholders in a smart parking solution.

City council administrators

These individuals will need to manage and monitor the smart parking solution, meaning they might require the following:

- A dashboard showing the occupancy level of the different parking spaces in the city
- An alert management solution to alert and notify council administrators of any issues that occur
- Analytics and reporting for smart parking solutions across the city
- A wholesale billing solution to help the council manage and bill different smart parking operators
- Data lakes and ML capabilities to enable council or government data scientists to run different ML predictive models to help with land planning and other logistics

Smart parking operators

Usually, councils do not run smart parking solutions themselves; rather, they make data such as parking space details and other data they can expose through a rich set of APIs available to third-party companies and partners to operate and manage the smart parking solutions.

The operators usually cover the following tasks:

- Setting up and converting parking spaces into smart parking spaces
- Offering vehicle owners different digital engagement channels, such as on web, mobile, or virtual assistant devices such as Alexa to do the following:

  a. Search for available parking lots nearby

  b. Book, reserve, and pay for an available parking lot

  c. Parking lot management, that is, updates, changes, cancelations, and so on

- Monitoring and operating the whole smart parking solution on a production-grade scale
- Exposing or selling smart parking APIs to other third parties and start-ups for the co-creation of different ecosystems for smart parking
- Integrating with internal and external capabilities such as payment gateways, OpenCV (OpenCV is an open source library for computer vision or image processing), and many other features as part of the final smart parking solution

Vehicle drivers

These are the end users and the target users at the end of the new business model or smart parking solution. Drivers will be provided with the features mentioned above by smart parking operators.

Smart parking solution details

Now let's think about how the smart parking operator will address the requirements we just discussed and come up with a solution.

IoT solution field planning

In networking, network engineers usually do a survey where network nodes such as switches and routers will be placed on the site, which kind of cables will be used, and which kind of network topology will be used (that is, star, mesh, ring).

Cables going through specific routes and even walls is not a haphazard task; there are engineering practices and explanations behind all of that setup.

IoT solutions are no different from traditional networking solutions. In fact, the IoT is sometimes classified as a networking engineering topic. It is called the internet of things, so connectivity and networking are a core part of it.

With any IoT solution, there is a physical world that it needs to sense and then act upon that sensing, which means in an IoT solution, there is always a field or playground where you need to deploy and install IoT devices and sensors.

Step number 1 in any IoT solution is to study very well the field or the ground where the IoT devices will be deployed. In our case, the parking space that we need to study is a field.

We need to know the following about the city's parking spaces:

- The number of parking spaces in the city and their location details
- Parking space type(s):

    A.   Is it a one-level parking lot or multi-level?

    B.   Is it off-street (a kind of closed/private parking space where there is a gate (in and out) controlling the vehicles coming in and out of the parking lot), or is it on-street (an open parking area) where there is no gate to control the vehicles coming in and out of the parking space?

- Parking lot type:

    A.   Within the parking lot, how many parking spaces are there? What type of parking spaces are they? That is, are they perpendicular, angled, or double parking spaces?

    B.   Any other metadata about the parking lot, for example, its latitude and longitude.

IoT solution devices and sensors

Given the above IoT solution field planning, the next steps will involve choosing the IoT devices and sensors needed to detect the occupancy status of the parking lot.

There are so many ways in which to detect the occupancy status of a parking lot. Here, we'll mention a few:

- You could install one or more – based on the parking space size – surveillance cameras to cover the whole parking space. Then you could use the ready-made, advanced AI Video Analytics technology to detect whether the parking lot has spaces or is occupied.

Previously, you'd have had to build such services from scratch, using technologies such as image processing and similar, but now, with AI and ML's maturity, hyperscale companies such as Amazon, Microsoft, and Google provide so many ready-to-use cognitive services, such as object detection, image classification, or activity detection (video), that you don't need to build an ML model yourself; it is already done for you. You might only need to feed the model with your data, that is, parking lot images or videos to show when they are occupied and when they are empty. I said might, as detecting a parking lot's occupancy status is becoming a standard, popular use case in that domain, so you might even find a third-party company that offers APIs ready to detect that. You would just need to pass to those APIs images of the parking lot you want to check the occupancy status of and you would get the answer: *Occupied* or *Empty*.

- You could install in each parking space an IoT device buried underground. That device would be equipped with an infrared sensor to detect vehicle motion and report accordingly whether a space is occupied. This option might be a bit expensive as you would have to install an endpoint IoT device with sensors underneath each parking space. Yes, we could have low-value ($5 or less), low-power (battery-based, lasting 10 years or more) IoT endpoint devices to serve that purpose, but the cost will be linear with the number of parking lots in all parking spaces across the city as a whole.

- You could install – in the case of gate-controlled parking spaces – a sensor on the in and out gate controllers to count the number of incoming vehicles minus the number of outgoing vehicles to detect how many parking spaces are available and how many are occupied. Detecting how many parking spaces are available might be enough in some cases, but in other cases, there might be a need to show the vehicle owner or driver where exactly the available parking space they booked or reserved is. For the latter use case, you need to use a different approach, as explained earlier.

Smart parking in offline mode (or IoT Edge)

You might, as part of smart parking business requirements, need to run some smart parking workloads at the edge, like the example we mentioned before about calculating the number of incoming vehicles versus the outgoing vehicles, or some real-time or near-real-time analytic requirements, and so on.

You must think carefully about such requirements. You need to think about the locations of where you will put the edge computing nodes. Edge units could be as simple as a small Raspberry Pi device or as medium complexity as a small-scale data center, or even as complex as a complete data center hosted at the edge.

In the case of off-street or private parking spaces, edge units could be placed in the parking spaces or nearby, while in the case of on-street parking, it would be a bit challenging to know where you should put the edge units. There are many options for that case, such as renting a nearby facility such as a building or even a room(s) in a shared building. You need, though, to think about the connectivity between the IoT endpoint devices on the field and such edge units, or you could buy managed edge services or hosting facilities from telecommunication companies or public cloud providers who have massive infrastructure footprints to host your units.

We have mentioned telecommunications companies as they have really massive footprint infrastructure coverage for a district or even – with 5G technology – street levels. That is because of their need to host telecommunication radio equipment. Public cloud providers also have edge locations but, for historical reasons, they don't have the level of coverage of telecommunication companies.

IoT solution connectivity

Given the preceding IoT solution field planning and based on the parking space type and details, you can, as a solution architect, define what kind of connectivity is required in the solution.

First, you need to ask yourself, I need to connect what to what? You basically need to connect the endpoint IoT devices to the internet or to the backend servers of the smart parking operators or council. Those backend servers are hosted in what is called the IoT Cloud or IoT backend cloud.

There are two options available:

- Connect endpoint IoT devices directly to the IoT backend cloud.
- Connect endpoint IoT devices to the edge gateway first and then the edge gateway will connect on behalf of those IoT endpoint devices to the IoT backend cloud.

With the first option, that is, directly connecting endpoint IoT devices to the IoT Cloud, you will need the IoT endpoint device to do the following:

- Have a communication module to support WAN connectivity, that is, like the LTE (LTE-M) module in the case of using a telecommunication network (a SIM card-based or Ethernet module in the case of using fixed broadband)
- Have an IP stack running in the device to support IP-based connectivity protocols

Those two conditions make such a connectivity option, that is, connecting to the IoT Cloud directly, not the preferred option due to the following:

- IoT endpoint devices – in the case of underground buried parking lot sensors – will become a problem as the cost will be high, that is, you have to have in each endpoint IoT device a cellular or LTE-M module with a SIM card. Also, in order to run the IP stack, the devices can't be constrained to low-value IoT devices.

- Those IoT endpoint devices will usually be on streets or far away from power supplies, hence they are usually battery-based devices. Running an IP stack on such endpoint IoT devices would impact the battery life of the devices, which usually work for 10 years or more to avoid the disturbance of changing such devices or their batteries.

The second connectivity option – that is, through the gateway – looks good as you could deploy low-value, low-power, and constrained endpoint IoT devices in the parking spaces. Those devices could communicate between each other and gateway devices using low-power local wireless networking such as Zigbee or Bluetooth Low Energy. Such connectivity technology will help in saving device batteries and it is cheap. The gateway will connect to the internet, meaning the gateway device could use a fixed broadband (wired option) if available or use cellular connectivity (wireless option). The gateway device or edge could or should be a powerful device in terms of computing resources.

We could use the following communication protocols:

- IoT endpoint devices could use the Zigbee protocol to communicate with other Zigbee-enabled devices, including the IoT gateway device.

- IoT edge devices or gateway devices – IP-enabled devices communicate with the IoT backend cloud using MQTT or HTTP(S) over the internet.

Smart parking IoT backend cloud

From the requirements, we will need to build (or buy) the following backend components for the smart parking solution:

- An IoT message broker/ IoT device gateway supporting MQTT, HTTP(S)

- A smart parking operational center – a web app for council and smart parking operators' admin users for managing and monitoring the smart parking solution

- Smart parking web and mobile apps – for the vehicle owners or drivers to book, manage, cancel, and pay for a parking space

- A smart parking API hub – an API hub for external third parties and partners looking to use the smart parking solution's data and its available set of APIs in their solutions

- A smart parking data lake – a data lake solution storing all smart parking data from all sources in its original raw format for further analysis by other downstream systems and solutions

- A smart parking analytic solution – an analytical and machine learning solution based on the smart parking data lake solution

Smart parking IoT solution – non-functional requirements

The features or functionality of the smart parking solution might be negotiated with business stakeholders. Usually, the most common statements you will hear at that stage are we can't boil the whole ocean, which feature is a must and which feature is nice to have?, and so on. There are lots of features requested by business stakeholders, but not all such features could be delivered in one release of the IoT project for a myriad of reasons, such as a lack of resources, the pressure of the delivery timeline, or even some features just being fancy features that the solution can work without.

But, when it comes to non-functional requirements, usually there's no negation with the assumption that the business stakeholders agreed with the increasing costs associated with achieving those non-functional requirements. Every business wants its solutions to be highly available (up and running 24x7), highly performant, highly scalable, highly reliable, and highly secure.

As a solution architect, you should consider non-functional requirements from the design stage and provide all the design principles, technical components, and controls required to fulfill such requirements. Also, those design principles should be applied from the device to the cloud, in each layer of the IoT solution.

The book's strategy

This section of this chapter is mainly to draw your attention to the philosophy or strategy that we will follow and use throughout this book:

- IoT ecosystems and technologies are so massive, and many technologies, solutions, and vendors are available for each IoT solution layer, from devices, sensors, and the edge to the IoT Cloud. Therefore, in this book, we will explain the concepts, and then we will explain the different options available. Finally, to be practical and pragmatic, we might go into the details of one, and only one, of a specific type of technology, solution, or vendor to show real examples where applicable.

- There are many IoT platforms available on the market, as we will explain in a later chapter. We will use one of those IoT platforms, which is the AWS IoT platform, and other AWS services that are related to building an E2E IoT solution. However, and as explained in the previous principle, the book will also touch upon other platforms if needed. For example, if a specific feature or service is not provided by AWS IoT and is provided by another platform, then we will cover that feature using the other IoT platform that provides the service.

- The book is basically for IoT solution architects and designers to design and build production-grade, large-scale IoT solutions, so we will not be able to go over some of the topics that will be covered in this book, such as connectivity, sensors, and devices, in-depth as they will require a dedicated book(s) in order to be covered in full. Having said that, the book will explain the concepts and some details that an IoT architect or designer should know about and understand when designing a complete E2E production-grade, large-scale IoT solution.

  For example, when we discuss the IoT cellular connectivity option, we will not delve deep into how it works in terms of radio access networks, telecommunication core networks, and so on. We will instead explain the concepts to some degree so that you understand at a high level how such types of connectivity work and how you can use them as part of your IoT solution blueprint.

Summary

In this chapter, we have learned what the IoT is, why we need the IoT, and what IoT solution architecture looks like at a high level. We covered different IoT solution architecture layers, starting with the IoT devices layer, which hosts IoT endpoint devices, sensors, actuators, microcontrollers, and so on. We also discussed the IoT Edge layer, which hosts IoT Edge computing and platforms, and then the IoT backend cloud, which hosts all IoT applications, platforms, and solutions. We explored the IoT connectivity layer, which is responsible for IoT connectivity across the whole IoT solution, and finally, the IoT security and access management layer, which is responsible for the security controls and access management of the whole IoT solution. We also covered different IoT solution design patterns, including Telemetry, Command, Bootstrapping, Device Twin/Shadow, and IoT device outbound connectivity. At the end of the chapter, we talked about a smart parking case study, which we will use throughout the book's chapters when needed.

In the next chapter, we will start with the anatomy of the word IoT. We will start with the letter *I* in IoT, which refers to the internet or, in other words, IoT connectivity.

2
The "I" in IoT – IoT Connectivity

Connectivity plays an important role in the **internet of things** (**IoT**)—without connectivity, it would not be called IoT, being called something else.

In this chapter, we will cover different topics around IoT connectivity. There is a huge number of different IoT connectivity technologies available on the market, hence we thought that it would be difficult and boring to delve deep into each connectivity option available. We followed another approach to tackle this challenge—we will start by exploring some of the important concepts regarding networking and connectivity in general, such as network topologies, the **Open Systems Interconnection** (**OSI**) model, the **Transmission Control Protocol/Internet Protocol** (**TCP/IP**) model, radio, wired versus wireless networks, network performance metrics, and many other concepts that are required to understand different IoT connectivity options. We will also look at the different factors that drive or enforce the selection of one specific IoT connectivity option (or a number of options) in IoT solution architecture. We will then move on to discuss these concepts in detail.

In this chapter, we will cover the following topics:

- Connectivity concepts
- IoT connectivity selection criteria
- IoT connectivity scenarios
- IoT connectivity options

Connectivity concepts

Let's go through a list of different concepts that will help us explain the different IoT connectivity options that we will discuss in this chapter.

Basic network components

If you have two computing devices or nodes such as a computer, printer, or server and they need to talk to each other, you simply connect those devices directly, either by a wire/cable or wirelessly (through radio waves, as we will explain later in this chapter). Each device should have a **network interface card** (**NIC**) or a modem that will be responsible for sending data (in the case of a sender device) or receiving data (in the case of a receiver device). Before sending data that is in binary format (that is, 0s and 1s) via a transmission medium such as cables (such as twisted-pair cables, coaxial cables, and optical fiber cables) or through a wireless medium such as radio waves, you will need to **encode** or convert the data into a format that is understood by the transmission medium. For example, if the transmission medium is a wire/cable, then the data has to take the form of electric voltage pulse signals (or pulses of light, in the case of fiber cables) that travel in the wire till they reach their destination. On the destination or receiver side, the opposite of that encoding process has to be done—for example, the electric pulse signals will return to their original form (that is, binary data) in order to be understood by the receiving computing node. This reverse process is called **decoding**.

In the case of wireless connectivity, the transmission medium is the space or the air, so the encoding or the radio modulation process at the sender node will convert the binary data into radio waves that traverse through the air or the radio spectrum until they reach the receiver node, which does the opposite and converts the radio waves into digital or binary data. Such an encoding/decoding process is managed and done by an NIC or a network modem.

The concept of encoding and decoding is very important in understanding networking, especially to understand how the last mile or the physical layer of the connectivity is done.

A node in the network should also have an **operating system** (**OS**) or an embedded system responsible for running the network software stack to be able to communicate with other nodes in the network—for example, running the TCP/IP stack and the User Datagram Protocol (UDP)/IP stack.

If you want to connect more devices to each other, then you will need a network component called a **switch** to facilitate that—a switch connects multiple devices to each other to create a local network.

Data is generated by the computing devices in the form of data packets—the **data packet** structure contains the data or the payload that needs to be transmitted, plus some other controlling data or headers such as the source physical address and the destination address. Switches (and routers—see the next component) use such control data to route the data packets to the right destination in the network. In fact, there are three types of data structures or **protocol data units** (**PDUs**) used for holding the data across the network: a **segment**, a **packet**, and a **frame**. We will touch on PDUs when talking about the OSI model later.

A **router** is another network component that is responsible for connecting multiple switches and, accordingly, their networks to form a larger network. In other words, switches connect devices inside the **local area network** (**LAN**), and routers are at the edge of the LAN and are responsible for connecting that LAN to another external LAN by talking to the other network edge router.

There is also an old network component called a **hub** that has now been replaced by a **switch** as switches are more advanced and smarter than hubs. A hub was previously used to broadcast data messages to all devices in the network, and then it was up to the receiving device to ignore or accept such data, while a switch sends the data message only to the right destination.

To give an example, in your home, the internet wireless **router** you got from your internet broadband provider usually does multiple networking functions. For example, it acts as a switch to connect home devices to each other; it also acts as a router to connect your home's local network to an external network (that is, the internet). Furthermore, a router also acts as a wireless **access point** (**AP**) for the wireless connections in the home.

The following diagram summarizes the basic network components we just discussed:

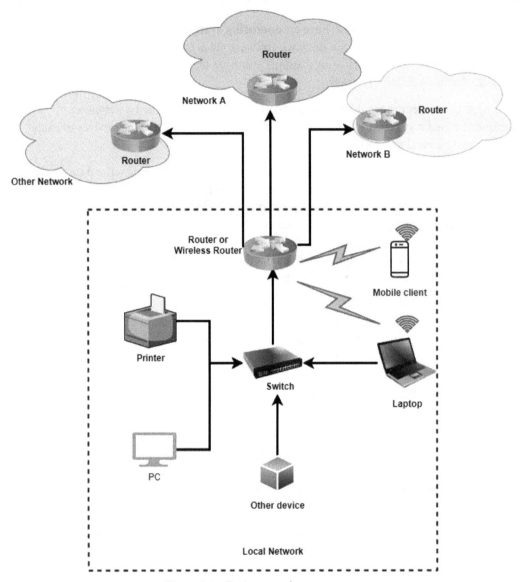

Figure 2.1 – Basic network components

Finally, if the devices want to talk/connect, then they need a language or a set of rules to use in their communication—that language is called a **protocol**. There are several protocols in different network layers. For example, in layer 7, or the application layer, you have **HyperText Transfer Protocol (HTTP)**, **HTTP Secure (HTTPS)**, **Simple Mail Transfer Protocol (SMTP)**, **File Transfer Protocol (FTP)**, **Domain Name System (DNS)**, **Constrained Application Protocol (CoAP)**, **Message Queuing Telemetry Transport (MQTT)**, and many more application protocols.

Now, let's move on to another connectivity concept: transmission modes.

Transmission modes

Data is transferred from one device to another through what is called a transmission mode, which defines the direction the data needs to travel in to reach its destination. There are three categories of transmission modes. Let's have a look at them next.

Simplex transmission mode

In this mode, the data is transmitted in one, and only one, direction, so the sending device can only send data but can't receive data, and the receiving device can only accept or receive data but can't send data. For example, in a television (TV) network, our TVs can only receive broadcast signals sent by the TV network.

This transmission mode presents no traffic issues as the data goes in one direction.

Half-duplex transmission mode

In this mode, the data can be transferred in both directions but not at the same time. For example, in radio devices such as a walkie-talkie, the receiver of the data (the data here is the voice) must wait until the sender of the data finishes replying to them. Usually, they say things such as Over to indicate that the sender has finished what they want to communicate to give the other side the chance to reply.

This model is good compared to the simplex transmission mode as you have two devices communicating with each other using the same channel. This is good resource utilization, but since they are not communicating at the same time, that adds one disadvantage: delaying or having high latency in the transmission.

Full-duplex mode

In this mode, both devices communicate by sending and receiving data at the same time. For example, mobile phones use full-duplex transmission mode. In this mode, there's no delay in the communication.

Now, let's talk about the famous network OSI and TCP/IP models.

OSI and TCP/IP models

The OSI model describes the seven layers that computer systems use to communicate over a network. The modern internet is based broadly on a simpler communication model called the **TCP/IP** model; the OSI model is still used, though, but not like the TCP/IP model.

In the following diagram, you can see the different layers of the two models:

Figure 2.2 – OSI and TCP/IP models

The main objective of the OSI or TCP/IP model is to outline and define how the data will be moved or transferred from one node to another node through the network. In each layer of those models, there are lots of protocols, standards, and technologies that help in achieving that goal.

The good thing about the layering or decoupling of system functions is it gives room for future development and adaptations; for example, when **IP version 6 (IPv6)**—which is part of the networking layer—was introduced, only the networking layer was impacted and other layers were not impacted.

An IoT network is not different from a traditional internet network—it kind of piggybacks on the existing internet network infrastructure and standards such as the TCP/IP model. In the preceding diagram, we showed the different protocols in each layer by focusing on commonly used protocols in IoT connectivity solutions.

Before we go through each layer, let's remember the main components that make up an IoT device or IoT sensing node, as follows:

- Microcontroller
- Sensors and actuators (that is, devices that can convert electrical signals into physical action, such as a relay for operating a power switch)
- **Real-time OS (RTOS)**
- Communication module(s)
- Embedded systems and **software development kits (SDKs)**

In this chapter, we will focus on the IoT device communication module. In IoT devices, you can have different types of communication modules. You could have an Ethernet communication module, such as the one shown in *Figure 2.3*, you could have a radio communication module to support different wireless connectivity options such as Wi-Fi, cellular, Zigbee, and Bluetooth or you could have both of those types of modules supported in the same device.

The **Raspberry Pi** device shown in the following screenshot (a kind of powerful IoT device) doesn't come with a built-in radio communication module. Thanks to the Raspberry Pi's pluggable architecture, you can buy add-ons such as a Wi-Fi dongle and plug it into the Raspberry Pi using one of the **Universal Serial Bus** (**USB**) ports available, or you can get that radio communication module as part of **Raspberry Pi HAT** (where **HAT** stands for **Hardware Attached on Top**), which acts as an additional card to Raspberry Pi and brings with it new features, such as coming with built-in sensors:

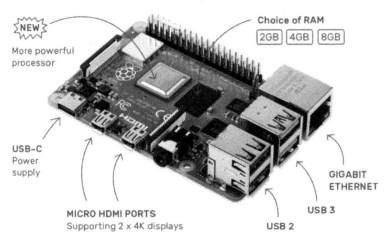

Figure 2.3 – Raspberry Pi 4 device (source: raspberrypi.orgraspberrypi.orgraspberrypi.org)

As another example, my 10-year-old daughter's micro-bit device shown in the following screenshot has a Bluetooth connectivity option (radio). See the red rectangle—it uses **Bluetooth Low Energy** (**BLE**), which we will explain later in this chapter:

Figure 2.4 – Micro-bit device with a radio communication module (BLE)

The devices we just discussed are only examples or an introduction to talking about the IoT communication module—we will deep dive into IoT devices and microcontrollers in the next chapter.

Now, let's go briefly through the TCP/IP model—we chose this model as it is simpler than the OSI model since it only has four layers. Let's try to understand how this model fits with IoT connectivity.

Application layer

This is the top layer of the TCP/IP model—it's the layer where the end user interacts with different applications. There are many protocols used in that layer—here, we mention a few of them:

- **HTTP/S**: This allows the end user to interact with web applications through a web client such as a browser. The browser acts as a client that sends messages to the web servers using the HTTP or HTTPS protocol.

- **FTP**: This protocol is used for transferring files between different nodes.

- **CoAP**: This application protocol is mainly for constrained devices and is commonly used in IoT applications.

- **MQTT**: This is a lightweight protocol that follows the **publish/subscribe** (**pub/sub**) model used to transport messages between devices. It is commonly used in IoT applications, and we will cover this in more detail later in the book.

And there are many other application protocols that already exist—or are yet to come—that work in this layer.

This layer runs on top of the transport layer—in other words, applications in that layer will generate data in a high-level format such as **JavaScript Object Notation** (**JSON**) or **Extensible Markup Language** (**XML**). Then, they will send that data as mandated per the selected application protocol. In that application layer, the data is pushed as a whole unit to the next layer—that is, the transport layer. The transport layer will break the whole unit of the application data into smaller parts called segments (in the case of the TCP protocol used in the transport layer) or datagrams (in the case of the UDP protocol used in the transport layer). A segment contains or holds two types of data—one part holds what is called control plane data or a transport control header, and the second part is the actual data or the payload that needs to be transmitted. The control header size varies between 20 bytes and 60 bytes, depending on the transport protocol used (TCP, UDP, and so on). That control header usually contains information such as the source port, destination port, sequence number, and checksum. Such information in the header helps in controlling the transportation of data to the destination node.

For example, the small data parts will be reunited again on the destination side based on the sequence number sent in each data part, the checksum will validate that all data has been sent successfully, and so on.

This process of splitting the data into small data parts is usually called fragmentation, and the rejoining of small data parts into the original whole part process is called a defragmentation process.

The data is usually split based on the defined network **maximum segment size** (**MSS**) attributes.

> **Important Note**
>
> In the application layer, the IoT applications—whether backend IoT applications or device IoT applications—are communicating using protocols such as MQTT, CoAP, HTTP/S, **Extensible Messaging and Presence Protocol (XMPP)**, and **Advanced Messaging Queuing Protocol (AMQP)**.

We will cover some of the IoT application protocols in more detail later in this book.

Transport layer

This layer is responsible for delivering messages from one node to another node in the network. There are several protocols in that layer, the most famous being UDP and TCP.

UDP protocol

This takes the data from the upper layer—that is, the application layer—and adds additional information to the data or the message: things such as the port addresses of the source node and destination node and the length of the data. The data construct in the UDP—as the name suggests—is called a datagram.

The UDP protocol is not as reliable as the TCP protocol as it is a kind of fire-and-forget protocol. The data will be sent to the recipient, but the sender will not wait for an acknowledgment to make sure the data has been received by the recipient, so there's no control over the traffic flow of the data messages between the sender and receiver. In other words, in UDP, there's no guarantee of message delivery.

UDP is much faster, lightweight, and more efficient than TCP, which is driven by the fact that the UDP protocol has less traffic control.

UDP is used in transporting messages of application protocols such as DNS, **Dynamic Host Configuration Protocol (DHCP)**, **Trivial FTP (TFTP)**, and **Voice over IP (VoIP)**.

TCP protocol

The TCP protocol is connection-oriented, meaning a TCP connection must be established between the sender and the receiver before the transmission of data. TCP takes the data from the upper layer—that is the application layer—and builds a new data construct— that is, breaking down the message into small chunks (segments), as explained before regarding the application layer.

TCP is used in transporting messages of application protocols such as HTTP, HTTPS, and FTP and is used as a transport protocol for IoT application protocols such as MQTT, HTTP, CoAP, and AMQP.

There are some other transport protocols implemented, either as part of specific technology, such as Zigbee (Zigbee has its full **end-to-end** (**E2E**) network stack, so it implements the transport layer as well and doesn't count on transport protocols such as TCP, UDP, or the transport protocol implemented to deal with a specific network) or through physical layers such as the LoRaWAN protocol, which was implemented by LoRa as its transportation protocol. We will discuss Zigbee and LoRa later in this chapter.

For now, let's move on to the next layer.

Networking layer

The main role of the network layer is the routing and forwarding of data packets from the sending node to the receiving node. The networking layer receives data in the form of data segments from the transport layer and creates a new protocol data unit called a **data packet**. The data packet contains the segments from the transport layer plus a new header, containing some information required in the routing in the networking layer. The information captured in the header of the packet contains information such as the source IP address, the destination IP address, **time to live** (**TTL**), and some other information.

This is the layer where the IP addresses of the sending and receiving nodes come into the picture. Those IP addresses are logical addresses, not physical addresses of the nodes.

The router is the main network component in that layer, and its main functions are outlined here:

- **Routing**: This is why it is called a router. When a packet reaches the router, the router inspects that packet and reads the information from the headers to decide the next hop or next path in the network to forward the traffic to.

- **Mapping between logical address and physical address**: The logical address is used to differentiate between source and destination nodes. In the network layer, the logical address (IP address) is used, but in the physical layer, the physical address (**media access control**, or **MAC**) is the address that will be used.

- Connecting different networks and forming what is known as a **wide area network (WAN)**.

- **Fragmentation**: This is a process of breaking packets into small chunks (frames) that then traverse through the physical layer. Frames are a new PDU containing the packet received from the network layer plus a new header containing information about the source and destination MAC addresses, checksum, and length. The source and destination MAC addresses will keep modifying as the frames traverse through the network, while routers modify the MAC addresses of the source and destination.

There are many protocols implemented in that networking layer, such as IPv4, IPv6, and **IPv6 Over Low-Power Wireless Personal Area Networks (6LoWPAN)**.

IoT IP-enabled devices can use IPv4 or IPv6; however, 6LoWPAN has been introduced to support low-power devices with limited processing capabilities to use internet protocols such as IPv6.

Physical layer

In this layer, the data messages from the networking layer are encoded in the data link layer frames and pushed to the communication medium, which can be either wired or wireless mediums. See the encoding/decoding concept we explained earlier.

Several protocols exist in this layer. Here are a number of them:

- Ethernet
- Ethernet **Time-Sensitive Networking (TSN)**
- BLE
- Wireless HART
- Zigbee
- Z-Wave
- **Radio-frequency identification (RFID)**
- **Institute of Electrical and Electronics Engineers (IEEE)** 802.11ah— Wi-Fi HaLow
- IEEE 802.15.4e
- LoRa/LoRaWAN
- Sigfox

- Cellular (**General Packet Radio Service (GPRS)**, **second-generation (2G)**
- **Third-generation (3G), fourth-generation (4G), fifth-generation (5G)**
- EnOcean
- **NarrowBand-IoT (NB-IoT)**
- **Worldwide Inoperability for Microwave Access (WiMAX)**

And there are many other protocols and standards in that layer.

We will explain more about those technologies and standards later in the chapter.

Now, let's talk next about another important concept in networking/connectivity: network topologies.

Network topologies

The way nodes or devices in a network are connected to each other is known as a **network topology**. There are five types of network topology, as detailed next.

Mesh topology

In a mesh topology, each device is connected to every other device on the network through a dedicated point-to-point link, as illustrated in the following screenshot:

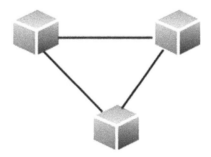

Figure 2.5 – Mesh topology

This kind of network setup is secure as it is a dedicated link between the connected nodes. It is also reliable as if one link fails, other links will not be impacted by that failure.

However, in a wired mesh network setup, the number of wires required will be a headache and a burden to maintain. Another issue is scalability—to add a new device, you need to connect it to all existing devices through dedicated links. In *Figure 2.5*, we just showed three nodes for the sake of simplicity in explaining this topology.

Star topology

In a star topology, each device in the network connects to a central device—that is, a switch. Devices in that topology do not connect directly to each other, unlike in a mesh topology. If a device wants to send data to another device, it will send it first to the switch, and then the switch will send or forward the data to the receiver device. The process is illustrated in the following screenshot:

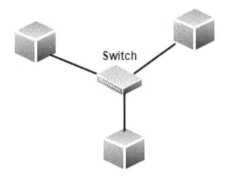

Figure 2.6 – Star topology

This kind of network topology is much easier to install and is robust. If one link breaks, the other links will not be impacted, but the problem with that setup is if the switch itself goes down, the whole network will be broken, as no device will be able to talk to any other devices in the network.

Bus topology

In a bus topology, each device in the network connects to a single cable called a backbone, and they then send the data through that cable to other devices. If the backbone is broken, then the entire network is broken. The process is illustrated in the following screenshot:

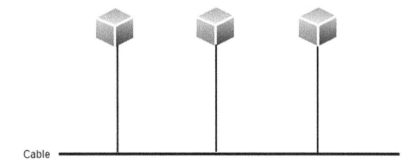

Figure 2.7 – Bus topology

Note that a bus topology is only used in wired networks.

Ring topology

In a ring topology, each device connects through a dedicated link to two devices on each side of it that will form a ring. If a device wants to send data to another device in the network, the sending device will send the data in one direction and the data will traverse through the ring. If a device receives data that is not meant for that device, then the device will forward the traffic to the next device in the ring, and so on until the data reaches the right destination device. The process is illustrated in the following screenshot:

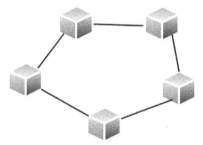

Figure 2.8 – Ring topology

This topology is easy to install and easy to scale too, but a problem may arise—if one link fails, the whole network will be broken.

Hybrid topology

As the name suggests, a hybrid topology means combining two or more network topologies together to create a new type of topology, as illustrated in the following screenshot:

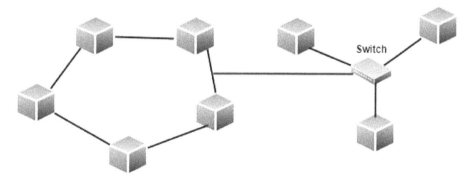

Figure 2.9 – Hybrid topology

This kind of topology is good for scalability and flexibility, but it is complex to set up and maintain. Examples are star-ring or star-bus hybrid networks.

Next, let's talk about the different network types.

Network types

There are many network types. Let's discuss them briefly.

LAN

In a LAN, devices are connected to each other in a small place such as a home, school, or hospital.

A LAN can be a wired or wireless network. If you have a wireless internet router in your home, that means you have a **wireless LAN** (**WLAN**).

Metropolitan area network (MAN)

This kind of network covers larger areas compared to a LAN—for example, it can cover a city or town. A MAN connects different LANs to form a bigger network called a MAN. It uses communication infrastructure—that is, telephone lines—to connect those LANs.

WAN

This kind of network covers larger areas compared to a MAN—for example, it covers a country, continent, or the whole world. The internet is classified as a WAN, and mobile 3G, 4G, and 5G mobile broadband is classified as a WAN as well.

PAN

This kind of network is formed around a person, so it connects computers, mobile, tablets, and digital assistants. It can be a wired PAN (such as a USB) or a wireless PAN (such as Zigbee or Bluetooth). The range of a PAN network is usually a few meters.

There are other network types such as a **home area network** (**HAN**), a **campus area network** (**CAN**), a **virtual private network** (**VPN**), and an **enterprise private network** (**EPN**), and so on.

Next, let's discuss the most important concept used in wireless technologies: radio.

Radio

Radio is the core technology behind all the wireless networks and devices around us today, so it is better to understand some basic radio concepts at a very high level without going into the academic low-level details, as this will help you in understanding the different wireless network connectivity options that we will discuss later in this chapter.

Definition of Radio – Source: Radio Systems Engineering, by Steven W. Ellingson

Radio is the use of unguided propagating electromagnetic fields in the frequency range of 3 **kilohertz (kHz)** and 300 **gigahertz (GHz)** to convey information. Propagating electromagnetic fields in this frequency range are more commonly known as radio waves. In a radio communication system, a transmitter converts information in the form of analog signals (for example, voice) or digital signals (that is, data) to a radio wave, the radio wave propagates to the receiver, and the receiver converts the signal represented by the radio wave back into its original form. Radio systems engineering encompasses a broad array of topics pertaining to the analysis and design of radio communications systems.

To explain further, let's talk about a basic radio setup as per the following diagram:

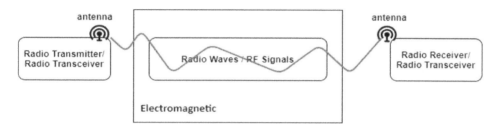

Figure 2.10 – Radio communication basic setup

In radio communication, there is a node that acts as a radio transmitter so that it generates radio waves that will traverse in the air as electromagnetic waves, and the radio receiver node from other ends will receive the radio waves.

Now, most radio chips support both functions (transmit and receive) at the same time, hence such a module is called a radio transceiver.

The idea now is simple—integrate that radio transceiver module in any device, and you will turn that device into a wireless device where the transceiver (in the case of sending data) will encode the message or the data into a series of radio frequency signals, and the transceiver on the other side (the receiver of the data) will decode the radio frequency signals into the expected format of the device that hosts that transceiver, as illustrated in the following diagram:

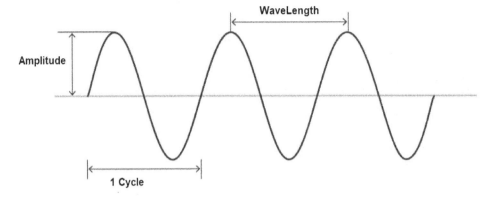

Figure 2.11 – Radio wave

Based on *Figure 2.11*, let's talk about some of the characteristics of radio waves and some other radio concepts.

Propagation

As with any other waves, radio waves propagate, hence they are used in transmitting data or information.

Amplitude

Amplitude is the highest point of the wave from the wave rest point.

Cycle

A cycle is a single change from up to down to up in the wave. 1 cycle in 1 second is measured by **hertz** (**Hz**), so if you have 1 cycle in 1 second, then you have 1 Hz; if you have 5 cycles in 1 second, then it is 5 Hz; and so on.

Frequency

Frequency is the number of events or cycles that occurred in a specified time interval. This is measured in Hz. As per the preceding cycle example, if we have 5 cycles in 1 second, that means the frequency is 5 Hz, so if we have 2,400,000,000000 or 5,000,000,000000 cycles in 1 second, that means we have 2,400,000,000/5,000,000,000000 Hz or 2.4/5 GHz. And guess what? These are the Wi-Fi standard frequencies that are used to transport data over the air to another radio device (radio receiver),), such as smartphones and tablets.

1,000 hertz is referred to as a kHz, 1 million hertz as a **megahertz** (**MHz**), and 1 billion hertz as a GHz.

High frequency means there are lots of cycles per second in the wave, and the longest propagation distance can be achieved by using a low frequency.

Wavelength

Wavelength is the distance between a point on one wave and the same point on the next wave. That point can be the crest of the wave (that is, the high point, as in *Figure 2.11*) or the trough of the wave (that is, the bottom point), so it doesn't matter which point you use to measure the wavelength.

Frequency band

This is a range of lower and higher frequencies in a frequency domain or service—for example, TV, mobile phones, and **Global Positioning System** (**GPS**) services use a frequency band called **ultra high frequency** (**UHF**), which covers a frequency range from 300 to 3000 MHz, and AM radio uses a frequency band called **low frequency** (**LF**), which ranges from 30 to 300 kHz.

Radio spectrum

There are many types of waves in electromagnetic waves. Electromagnetic waves that operate in frequencies ranging from 30 Hz to 300 GHz are called radio waves—such a range of electromagnetic waves is called a radio spectrum.

There are two categories of the radio spectrum—one is known as a licensed spectrum, which is a set of frequency bands regulated by national law and governed by governments.

Different parts of the licensed radio spectrum are allocated by the **International Telecommunication Union (ITU)** for the different radio transmission technologies and applications; usually, the licensed radio spectrum is sold to mobile operators or private radio services. The other category is an unlicensed spectrum, which is free for use by anyone but has one big issue because of such freedom: signal interference, as anyone can use that free spectrum in their radio service, so different applications using the same unlicensed radio frequencies will have their signals interfere with each other, making it an unreliable option. This is why mobile operators, TV services, and other private radio services use a licensed spectrum to avoid such interference.

To give an example of the drawback or the pain of an unlicensed radio spectrum, in my home, I have a digital shower that is controlled by a wireless shower controller device. Because I have some other wireless devices in the home, sometimes the wireless shower controller does not work until I have switched off other wireless devices or changed—if I can—their radio channels or frequencies as I suspect there is another wireless device that uses the same radio frequency that is used by that shower controller, hence signal interference occurs and the shower controller stops working accordingly.

So, manufacturers of such radio chips that use an unlicensed spectrum should handle such signal interference issues—for example, they can offer different radio channels or frequencies and let the consumer choose from those channels to avoid such interference with other existing radio services.

Radio modulation

We have now learned about some radio concepts, but the question remains: How cancan information be traversed through radio waves? The answer is by using radio modulation, which, in simple terms, is a way of encoding the data into a radio-wave form. There are two types of modulation: **amplitude modulation (AM)**, which uses amplitude to encode the information into radio waves, and **frequency modulation (FM)**, which is the most common radio modulation type used, which simply uses the frequency as a way of encoding messages.

Wired networks versus wireless networks

A wireless network uses radio signals to connect devices in the network, while a wired network uses wires or cables such as copper or fiber-optic cables to connect devices in the network.

A wireless network has many advantages, such as scalability—that is, you can add a new node easily to the network without impacting other nodes. Another great benefit of a wireless network is mobility, or freedom of movement. Wireless networks have some disadvantages, such as a limited range for radio signals, electromagnetic interference from nearby devices, and poor signal coverage in dead spots (that is, basements or underground places). And finally, there are security concerns with such networks as anyone can intercept the radio signals, hence it has to be encrypted.

A wired network has also some advantages, such as moving data faster than wireless networks. Also, there's no interference, such as electromagnetic interference, which may be experienced with a wireless network, and it is much more secure than a wireless network. The disadvantage of a wired network is a lack of mobility and expensive installation and setup.

Finally, the best setup is to use both types of networks. For example, cellular networks use a wireless network or a **radio access network** (**RAN**) as the air interface—that is, while you are walking in the street, for example, you communicate with the mobile operator base station using radio signals generated from your handset transceiver, but the base station itself might be connected to the mobile operator core network using a wired network. A similar setup exists in your home but on a small scale—you move freely inside your home with wireless devices such as phones and tablets. Those devices connect to a Wi-Fi router using radio signals, and then the Wi-Fi router is connected to the internet using old telephone lines or fiber-optic cables.

Next, let's discuss the most common network performance metrics.

Network performance metrics

Any network has the following metrics to measure its performance.

Bit rate

This refers to how many **bits per second** (**bps**) can be transmitted in the network. The measurement units are **kilobits per second** (**Kbps**) (1,000 bits per second), **megabits per second** (**Mbps**) (1,000,000 bits per second), **gigabits per second** (**Gbps**) (1,000,000,000 bits per second), and so on.

Throughput

The amount of data successfully transmitted in the network. Network throughput is measured in bps.

Latency

The amount of time taken to transmit data between sender and receiver nodes.

Packet loss

This is the percentage of data packets lost in communication concerning the number of packets sent.

Packet delay

The amount of time required to push the packet's bits into the transmission medium (wired or wireless). There are usually four sources of packet delay: processing, which is the time the router takes to process the data once it receives it; and queuing, transmission, and propagation, which is the time the bits take inside the medium to reach the destination.

For example, if the medium is a fiber-optic cable in the case of a wired network, then the propagation will be much faster if it is a copper cable, and so on.

Now that we have been through the network concepts and before going into the different IoT connectivity options, let's understand the factors or drivers you should consider when selecting a connectivity option(s) for your IoT solution.

IoT connectivity selection criteria

There are several IoT connectivity options we could use in our IoT solution, and selecting which connectivity option to use is driven by the following criteria.

Amount of data

How much data needs to be transferred from the IoT endpoint device or IoT edge device or to the IoT cloud? Is it in bytes, KB, and so on?

Frequency of sending data

How frequently do you need to send data from the IoT endpoint device or IoT edge device to the IoT cloud? Is it every second, minute, day, and so on?

Communication range (distance)

What is the distance between the source of data (IoT endpoint device and/or IoT edge device) and the destination (IoT cloud)? Is it a few meters, miles, and so on?

Communication latency (speed)

What is the speed required in transferring the data? Is it in milliseconds, seconds, minutes, and so on?

Device types

Is the IoT device battery-based (low-power), in which case we should be careful about the selected connectivity option to not select a connectivity option that eats batteries that are supposed to live for 10 years or more, or is there a source of power for IoT devices (high-power)? Technologies such as Powerline Ethernet are a good example of powered or plugged-in devices where data and electricity go together through the same cable or wire, or you could have a powered device that has a dedicated source of power and uses a different connectivity option (that is, wireless or wired separately).

The mobility status of the IoT device

Is the IoT device fixed in its location or is it movable? If it is movable over a large distance, such as with connected cars, then wired connectivity is not an option at all.

Now, before we cover the different IoT connectivity options, let's talk about some connectivity scenarios you might face in any IoT solution.

IoT connectivity scenarios

There are many connectivity scenarios, but we will focus on two common scenarios that exist in any IoT solution.

Have a look at the following diagram:

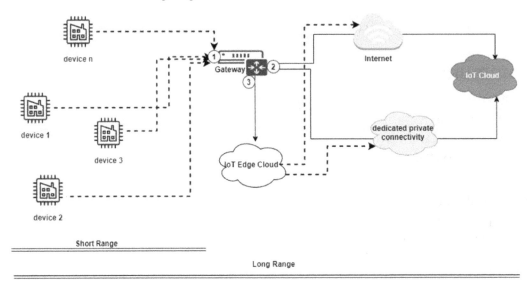

Figure 2.12 – IoT connectivity scenarios

As per *Figure 2.12*, we could have the following scenarios.

Scenario #1

In this scenario, you have the following:

- **IoT endpoint devices** (**device 1**, **device 2**… **device n**): They are battery-based (low-power) IoT devices, they are fixed in their location, they have a radio communication module for wireless communication, they are low-cost devices, they require a low data-bits rate on the communication link, and their radio module is manufactured to operate on an unlicensed radio spectrum.

- **IoT gateway device**: You have one or more of these; they are plugged in or have a source of power through an electricity source or some other power technologies such as energy-harvesting technologies. The IoT gateway has internet connectivity (interface number **2** in *Figure 2.12*). Such internet connectivity could be done through fixed broadband—that is, Ethernet (wired connectivity)—or mobile broadband or cellular connectivity (wireless connectivity). The communication module of the IoT gateway device in the case of cellular connectivity usually contains a mobile **subscriber identity module** (**SIM**) compartment besides other normal wireless or radio communication module compartments or components such as a transceiver or an antenna.

You get the SIM from the mobile network operator in the proper form (large, mini, micro, and so on), as manufactured by the communication module chip. Mobile network operators usually have a dedicated IoT network that differs from the mobile consumer network, as IoT network characteristics are different. The gateway device also has another radio communication module (interface number **1** in *Figure 2.12*) that supports radio technology used in the endpoint IoT devices to communicate with the endpoint IoT devices from one end and communicate externally through the internet connectivity with the IoT cloud from the other end. The gateway could also have or support multiple different radio technologies if the endpoint IoT devices use different radio technologies.

Now, based on the previous requirements, let's see the connectivity options we could use for these IoT endpoint (leaf) devices.

Connectivity option for IoT endpoint devices

IoT low-power endpoint devices should communicate or form what is called **LoWPAN** LoWPAN. There are several LoWPAN technologies, such as Zigbee, 6LoWPAN, WirelessHART, BLE, Z-Wave, and so on.

We will not cover the details of all those technologies, but as a rule of thumb, note the following points:

- They are all based on radio technology, they all operate in the unlicensed radio spectrum, and they are all short-range connectivity technologies. They follow the IEEE 802 standard, which is the big family of IEEE standards for LANs, PANs, and MANs. For each network option mentioned, there is a dedicated subgroup in that IEEE 802 family working and creating standards for those technologies. For example, the workgroup for Zigbee is IEEE 802.15.4, the workgroup for BLE is 802.15.1, and so on.

- The differences between those low-power PAN technologies are usually around the following technical features:

  - **Frequency band**: The Zigbee frequency band, for example, is 2.4 GHz (global), 915 MHz (Americas), and 868 MHz (Europe), while BLE operates at a 2.4 GHz **Industrial, Scientific, and Medical** (**ISM**) band (2.402-2.480 GHz utilized), and so on.

  - **Data rate**: Zigbee has raw data throughput rates of 250 kbs, which can be achieved at 2.4 GHz (16 channels), 10 kbs at 915-921 MHz (27 channels), and 100 kbs at 868 Mhz (63 channels), while BLE can transfer up to 2 Mbps.

- **Distance or communication range**: The Zigbee transmission distance range is from 10 to 100 meters, depending on power output and environmental characteristics. Sub GHz channels transmission ranges up to 1 **kilometer** (**km**); the BLE transmission range can transmit up to 100 **miles** (**m**) (Bluetooth 4.0 LE) and 400 m (Bluetooth 5 LE), and so on.

- **Network topology**: Zigbee supports a star, tree (star-bus network or hybrid), or mesh network topology, while BLE supports a star network topology.

Let's discuss examples of only two options (Zigbee and BLE) for the sake of simplicity and to give an idea about how such technologies work. We could, of course, list down from the IEEE standards all the technical features of each connectivity technology option we listed previously.

In a real IoT solution, you could end up choosing one or more connectivity options of those short-range LoWPAN connectivity technologies. There are many factors in choosing which connectivity option you could use for IoT endpoint devices, such as what kind of radio module/chipset is manufactured and supported in the endpoint devices.

You might end up or have to use a specific technology such as Zigbee, for example, if the solution IoT endpoint devices are supporting Zigbee only, and so on. Another factor is whether the IoT endpoint devices are high-power or low-power, and finally, we need to consider the technical features we mentioned previously.

Important Note

Those connectivity technologies mainly work and operate in the physical/data link layer of OSI or the TCP/IP model—in other words, they have protocols in transmitting data through the communication medium. In the case of LoWPAN connectivity options, the medium is the radio spectrum, but some technologies such as Zigbee provide an application and network layer beside the physical layer as well.

To solidify the understanding of that important point, in the following diagram, we have put the TCP/IP model on the left side and the Zigbee model on the right side:

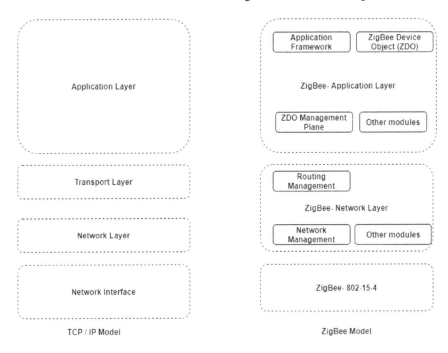

Figure 2.13 – Zigbee stack

As you can see, Zigbee has a full E2E stack; it has its own transport, network, and application layers on top of a standard implementation of the physical layer (IEEE 802-15-4). In other words, Zigbee applications dodo not use application protocols such as HTTP, MQTT, and CoAP, but use the Zigbee application protocol. And Zigbee applications dodo not push data to transport layers such as TCP and UDP but push it to the Zigbee transport layer. Then, the Zigbee transport layer does not push the data to a network layer such as IP but pushes it to the Zigbee network protocol. And finally, the Zigbee network protocol pushes the data down to the Zigbee physical layer, which implements the IEEE 802-15-4 standard (that is, data is transmitted through radio waves).

Zigbee provides a full E2E stack from the application to the physical layer. From a development point of view, there are several development libraries in different programming languages such as Java, Python, and others implementing the Zigbee stack. Here are a few of them: **zigpy** (https://github.com/zigpy/zigpy), **zigpy-xbee** (https://github.com/zigpy/zigpy-xbee), **zigpy-deconz** (https://github.com/zigpy/zigpy-deconz), and so on.

In the next section, we will briefly explain these short-range wireless technologies. Now, let's cover the connectivity options we could use for IoT gateway devices.

Connectivity options for an IoT gateway device

For IoT gateway device connectivity, we need to consider the following points:

- Does the IoT device gateway need to relay or forward the IoT endpoint device-sensed data to the IoT cloud, which is usually hosted far away from the IoT field, or does it just need to relay the IoT endpoint device-sensed data to the IoT edge cloud, which is usually hosted near the IoT field where the IoT endpoint devices are deployed, and then the IoT edge cloud itself connects to the IoT cloud?

- If the IoT gateway will just relay the data to the IoT edge cloud, then connectivity options could be wired or wireless short-range. If the edge is far away from the IoT field, then long-range wired or wireless connectivity options should be used.

We have already covered short-range connectivity options with IoT endpoint devices. Now, what about long-range connectivity options? We have two categories for long-range connectivity options, as outlined here:

- **Wired option**: In this case, the IoT gateway or the IoT edge cloud is connected by Ethernet toto the IoT cloud through either internet (public WAN) or through a dedicated private link or lease line (private WAN). Note that we are not discussing security here in this chapter; otherwise, we could discuss having **IP Security** (**IPsec**) or a VPN over such an internet or private connection, and so on.

- **Wireless option**: In this case, the IoT gateway or the IoT edge cloud is connected by wireless long-range technology such as the following:

  - Licensed cellular mobile broadband 3G, 4G, Long-Term Evolution (LTE), 5G, and satellite: Those options are valid with high-power devices; for low-power devices, you could use licensed cellular mobile **low-power WAN** (**LPWAN**) technologies such as NB-IoT or **Long Term Evolution for Machines** (**LTE-M**).

  - Unlicensed LPWAN technologies such as LoRaWAN and Sigfox for low-power devices.

To better understand the long-range wireless connectivity option, we could apply the same technical characteristics used in short-range connectivity options, such as the following:

- **Radio spectrum**: LoRa and Sigfox use an unlicensed radio spectrum, which means that interference could happen as they use a free and public radio spectrum to run their network. From a cost perspective, LoRa and Sigfox are cheaper compared to cellular technologies, but licensed cellular technologies provide more reliability (no interference), security, and more availability (the footprint of mobile LTE networks is more than LoRa networks and Sigfox networks).

- **Data rates** (uplink and downlink): LTE LPWAN technologies such as NB-IoT and LTE-MM provide more bandwidth and throughputs compared to LoRaWAN and Sigfox.

- **Distance or range**: LTE LPWAN technologies such as NB-IoT and LTE-MM provide more distance and coverage compared to LoRaWAN and Sigfox.

Now, let's think about another scenario we could have in an IoT project.

Scenario #2

In this scenario, the endpoint IoT devices are powerful, with a fixed source of power, or they are battery-based, IP-supported devices and just want to connect directly to the IoT cloud, which is far away from those IoT endpoint devices.

In that scenario, there's no need for short-range connectivity options, and the focus will be on long-range connectivity options, which we already discussed in *Scenario #1*.

Important Note

Scenario #1 is the most common scenario used in IoT solutions and the one that has features you can use, such as analytics @Edge. It is a cheap option as it uses cheap (free) connectivity options with a massive number of endpoint devices (you don't need to buy a cellular SIM for each endpoint IoT device in the project; you could use one with endpoint IoT device technologies such as Zigbee, which is very cheap, and then with 1 or 2 or even 10 or more gateway devices that use cellular connectivity to get internet access).

If you think about *Scenario#1*, it is similar to your home internet setup. You have wireless devices that use Wi-Fi radio access technology; you have a wireless router. That router is connected to the internet or has internet access, and also it acts as a wireless AP or home base station. All Wi-Fi-supported devices connect to it, data is transferred using Wi-Fi protocols, and then the router will receive the data coming through radio waves from devices and convert it or encode it into an Ethernet-expected format—that is, bits to be pushed into the wire, which, in this case, based on a home setting, is telephone circuits, and so on.

Now, let's talk briefly about different IoT connectivity options.

IoT connectivity options

In the previous section, we covered lots of concepts about networking, wireless technologies, and different IoT connectivity needs, which were needed to understand the different IoT connectivity options we will mention in this section. We will not go into the low-level details of such technologies, but rather, will give a high-level overview of them.

We can classify connectivity broadly into two categories: **wired** and **wireless** connectivity. For wired connectivity, we have the options outlined next.

Classic Ethernet (IEEE 802.3)

The Ethernet protocol is a well-known and widely adapted internet protocol used in transmitting data at high speed on the LAN. This protocol covers the physical and data link layer of the OSI model. There are different kinds of Ethernet standards, and new standards provide data transfer at a rate of 10 Gbps.

The physical layer consists of cabling and devices; there are different Ethernet cables such as coaxial cables (a bit old), twisted-pair cables, and fiber-optic cables.

The most used cables are twisted-pair cables. These have different categories, such as CAT 5/5e (data transfer rate of 100 Mbps), CAT 6 (up to 1 Gbps), CAT 6a (up to 10 Gbps), and CAT 7 (up to 10 Gbps). At either end of twisted cables, there are Rj-45 8-pin connectors responsible for sending or receiving data in either half-duplex or full-duplex mode.

There are several benefits of a wired Ethernet network. We mention a few of them here:

- It is adapted worldwide and is a completely open and free standard.

- It provides a much faster speed compared to wireless networks.

- There is no interference as there is with wireless networks, and it is more reliable compared to a wireless network.

- An Ethernet cable can be up to 100 meters—you could say this is the Ethernet network range.

- There's no security overhead—that is, username/password in case of Wi-Fi or pairing in the case of Bluetooth. With Ethernet, you can just plug it in and you are ready to transmit the data.

- It is not expensive.

The clear drawback of a wired network that leads to a wireless network is the mobility and movement of network devices. IoT solutions commonly use wireless technology, but that doesn't mean Ethernet is not used in IoT—it is used, but not more than wireless networks and technologies. Here are some examples of IoT devices connected using Ethernet:

- Surveillance camera and video

- VoIP

- Set-top boxes for video and audio streaming

- Industrial high-reliability devices such as robotics utilities, transportation, medical, and aerospace devices—see the *Ethernet TSN* section

- Game consoles

Ethernet TSN

Ethernet TSN is a new standard on top of the Ethernet standard protocol, aimed at industrial networks with automation and controls. For applications that need a deterministic network that works in a predictable fashion, Ethernet TSN introduces the following benefits:

- **Security**: Ethernet TSN increases security as it uses standard Ethernet.

- **Scalability**: Ethernet TSN can be used for 100 Mbps as well as for 1 Gbps or 5 Gbps.

- **Open standard**: Ethernet TSN is managed by the IEEE 802.1 standards committee. Many technology providers support Ethernet TSN.

- **Low cost**: Ethernet TSN leverages existing Ethernet technology, so it is not expensive.

Now, let's explore briefly the different wireless networks commonly used in IoT solutions.

We can categorize wireless networks into two categories: **short-range** wireless networks—
that is, the range or the distance of radio waves or frequency is very short or limited—
and **long-range** wireless networks—that is, the range or the distance of radio waves or
frequency is medium to long.

Let's start by exploring short-range wireless networks.

Wi-Fi

Wi-Fi is the most common wireless network technology used in the consumer electronic
sector and in enterprises too. You will find Wi-Fi in homes, offices, trains, shopping malls,
and many more places.

Wi-Fi is based on the IEEE 802.11 standard, which is part of the IEEE 802 standards
(LAN standards), so IEEE 802.11 covers wireless local network standards. IEEE 802.11
standards have gone through multiple iterations or amendments (IEEE 802.11b, IEEE
802.11g, IEEE 802.11n, and several other iterations with alphabetical letters after IEEE
802.11), with each iteration offering, supporting, or enhancing the data transmission rate
or the range of communication (indoor and outdoor).

Regarding the OSI model, IEEE 802.11 standards cover the physical and data link layer, so
they define the standards and protocols that technology vendors implement in layer 1 and
layer 2 of the wireless LAN.

Wi-Fi can be used in WPANs as well as in WLANs.

Wi-Fi networks are secure, reliable, convenient, and fast, and have a similar level of
performance as wired Ethernet networks. This is why, at home, for example, you connect
your laptop using Wi-Fi despite the availability of Ethernet, which you could also use,
but you would require an Ethernet cable to be connected to the router, which is not a
convenient option.

Wi-Fi networks operate broadly in the 2.4 GHz and 5 GHz radio frequency bands. Some
other radio bands can be used as well, such as 6 GHz and 60 GHz. You might have Wi-Fi
devices operating on 2.4 GHz only or 5 GHz only, or both (dual band).

GHz and 5 GHz and other Wi-Fi radio frequency bands are on an unlicensed radio
spectrum, which means the following for Wi-Fi networks:

- The data transmission cost is low in Wi-Fi networks. (This is why most Wi-Fi
  networks are free or have a low cost. The Wi-Fi network in your home is free—you
  pay mainly for the **internet service provider's (ISP's)** cost of providing internet.
  The ISP connects the Wi-Fi router with the internet.)

- Signal interference could happen with other wireless devices that use or operate on the same radio band—that is, 2.4 GHz/5 GHz.

- Such high frequencies will require more power, which makes it a bit difficult for IoT devices/microcontrollers, which are (or require) low power and are low-cost.

Wi-Fi commonly uses the star network topology, whereby the AP acts as a switch, hub, or router. All Wi-Fi devices connect to it rather than directly connecting to each other.

There's also another network topology used in Wi-Fi networks, which is a point-to-point topology where devices communicate with each other directly. In a star topology, you will have the router act as a base station or receiver for radio frequencies from one end and connect to the internet—if needed—from the other end, while internet connectivity in the case of a point-to-point topology will be tricky or expensive, as each device has to have internet connectivity without going through the central gateway for internet access.

One of the biggest threats in any wireless network is security, especially with unlicensed spectrum technologies such as Wi-Fi, but the good thing about Wi-Fi networks is they have been built with security in mind from the beginning. Wi-Fi networks secure or authenticate who can access the network and encrypt data transmitted wirelessly in the network. Yes—it is not an easy task to set up security and authentication for a massive fleet of IoT devices that use Wi-Fi as a connectivity option, but this is another topic that can be discussed and covered with something such as IoT device management solutions, which we will cover later in the book.

> **Important Note**
> A Wi-Fi stack is widely used and implemented in many different OSs and embedded systems.

Now, let's move on to another interesting version of Wi-Fi connectivity, which fits in well with IoT.

Wi-Fi HaLow

If you think about the majority of IoT application use cases, you will come to the following conclusions:

- **A low data rate is required**: For example, in smart meter use cases, you need every month, week, or so to transmit a small amount of data—that is, meter readings that can be just a small byte of data.

   The same also applies to smart parking use cases—you just need to know whether a parking lot is occupied or not, and so on.

- **Deep network coverage is required**: There are what is known as dead-spot locations where wireless or radio signals can't reach or can't penetrate the walls and other obstacles to reach such locations—for example, a smart meter (or smart parking lot device) is buried in the basement or underground, so in order to reach such places, you need to use a low radio frequency (sub 1 GHz), not a high radio frequency.

  As explained in the Connectivity concepts section, a low frequency has good penetration and network coverage.

- **Low power is required**: Using the same example as before, you might have smart meter devices (or smart parking lot devices) deployed in the basement or underground where there's no source of power, hence they are battery-based. As the operator of those devices, you need to save their battery to keep them alive for at least 10 years or more to avoid the operational burden of changing such device batteries frequently. If we are talking about large-scale smart meters deployed in a city, for example, then it would be a real operational burden to change the device batteries at that scale.

The preceding three factors are what led the Wi-Fi Alliance group to support the new standard of low-power Wi-Fi (IEEE 802.11ah—Wi-Fi HaLow) wireless networking standard that uses the 900 MHz (below 1 GHz) license-exempt band to offer long-range (approximately 1 km) and low-power connectivity. In other words, it has been built mainly for IoT.

Bluetooth classic

Bluetooth connectivity is like Wi-Fi connectivity in terms of popularity; it is a worldwide known connectivity technology used in lots of consumer electronic devices such as wireless headphones, keyboards, mouses, phones, laptops, tablets, and many other consumer electronic devices.

Bluetooth operates on an unlicensed industrial, scientific, and medical (ISM) band at 2.4 GHz.

Bluetooth offers short-distance (the range varies from 1 m to 100 m) wireless communication with low power consumption (Bluetooth devices can sustain battery life for weeks or months).

The data transfer rate in Bluetooth can be up to 2 Mbps. Network topologies used in Bluetooth are star, point-to-point (also known as piconets), and—recently—mesh topologies, which add lots of benefits, such as extending the range of communication by using intermediary nodes to relay data throughout the network.

Bluetooth provides security and access control; in the pairing process, devices authenticate to each other, and data is transmitted wirelessly in an encrypted format.

Bluetooth stacks are supported on multiple OSs such as Windows, macOS, iOS, and Android. In Linux distributions, there is BlueZ, and in embedded systems, there is Zephyr.

BLE

Bluetooth Low Energy is part of the LoWPAN family; it has the same story (regarding classic Bluetooth) as the story of classic Wi-Fi (regarding Wi-Fi HaLow). The BLE protocol was introduced to offer low-power and long-range Bluetooth connectivity to serve different IoT applications' needs.

It leverages the same frequency band as classic Bluetooth—that is, a 2.4 GHz ISM band—but it uses different channels: it uses 40 channels with 2 MHz spacing, while classic Bluetooth uses 79 channels with 1 MHz spacing.

Zigbee

Zigbee is part of the LoWPAN family, so it is a low-power radio technology, its standard and protocols are free, and it has been adopted by many chips and device manufacturers. Zigbee is supported and maintained by the Zigbee Alliance entity.

Zigbee was built based on IEEE 802.15.4 (**LR-WPANs**), the standard that defines the physical and data link layer for LR-WPANs.

Zigbee operates on an unlicensed band, both regional and worldwide. It operates at 2.4 GHz worldwide; in that worldwide option, it has the same frequencies as Wi-Fi and Bluetooth. It operates at 915 MHz in the **United States** (**US**) and at 868 MHz in Europe.

It is a perfect choice for low-power, low-cost IoT endpoint devices; it is low-bit-rate technology; it supports data rates at 250 Kbps, 100 Kbps, 40 Kbps, and 20 Kbps. The range of Zigbee radio is approximately between 10 and 100 meters.

It supports different network topologies such as star, mesh, and tree (star-bus), and the Zigbee network can support up to 64,000 nodes.

Zigbee provides network security management and encryption of data transmitted throughout the Zigbee network.

Zigbee provides a complete stack, as explained earlier in this chapter; in other words, it is not networking technology that covers the physical and data link (or the **medium access controller**, or **MAC**) layer only. It also covers the network layer and the application layer. If you compare it with Wi-Fi and Bluetooth, you will notice that Wi-Fi and Bluetooth cover only the physical and data link layers, and they count on other protocols to cover the other layers. For example, if you wanted to run an HTTP-based IoT application or an MQTT-based IoT application, you'd use HTTP and MQTT protocols in the application layer, then you'd use the TCP protocol for the transmission layer, then IP for the routing layer, and then Wi-Fi or Bluetooth for the data link and physical layer. However, in Zigbee's case, it is E2E from the application to the physical layer, which is done by the Zigbee stack.

In a typical Zigbee network architecture, you would find the following set of devices or nodes:

- **Zigbee end node**: As the name suggests, these are the nodes or the sensed nodes that need to send and receive data using Zigbee connectivity.

- **Zigbee router node**: As the name also suggests, this passes data between the Zigbee network nodes.

- **Zigbee coordinator device**: This node communicates with the Zigbee router and other Zigbee nodes.

- **Zigbee Trust Center node**: This is the node responsible for configuring and authenticating Zigbee nodes such as routers and end nodes.

Zigbee is a very cheap option in comparison to Wi-Fi. To give an example, inside Philips Hue bulbs or IKEA smart bulbs, there's a Zigbee chipset that does both tasks of connectivity and LED control. The only challenge with Zigbee in the case of smart home solutions is that it is not connected to the internet; to connect such smart bulbs to the internet to control them remotely through a mobile or web app, you need a Zigbee gateway to enable that.

Now, let's discuss another alternative option to Zigbee, which is Z-Wave.

Z-Wave

Z-Wave is another low-power wireless communication protocol that operates at an unlicensed spectrum at 800-900 MHz. Because it operates at a sub-1 GHz radio frequency, it can penetrate walls and concrete better than high frequencies such as 2.4 GHz, which is used by Zigbee, Wi-Fi, and Bluetooth, which means the Z-Wave range has a little more compared to the Zigbee range.

Z-Wave is not open like Zigbee—it is a proprietary protocol developed, managed, and maintained by a company called Zensys, which is one of the reasons why Zigbee is preferable to Z-Wave. There are other reasons, such as cost, as Zigbee is cheaper than Z-Wave due to the proprietary licenses.

The Z-Wave data rate is up to 100 Kbps, the range is up to 100 m, and it uses a mesh network topology.

It is used mainly in home automation solutions and it is like Zigbee; it needs a Z-Wave gateway to get internet connectivity.

NFC

Near-field communication (**NFC**) is a set of wireless connectivity protocols and standards to enable communication between two devices in a range of 4 centimeters (cm) or less.

NFC is popular in consumer electronic devices and applications. If your phone supports NFC, you can have applications such as eWallet or mobile payments in general. It is also used for contactless cards, electronic tickets, and in many other use cases.

In IoT solutions, NFC might not fit due to the very short range it operates with—that is, 4 cm—but it can be used as a bootstrap or provisioning step to endpoint IoT devices.

> **Important Note**
> The preceding connectivity options are the most common short-range connectivity options in any IoT solution. We covered both high-power and low-power technologies, but there are other options such as 6LoWPAN, for example. We didn't mention this just for the sake of simplicity, but the concept is the same.

Now, let's see a list of IoT long-range wireless connectivity technologies.

Cellular connectivity – 3G/4G/LTE/5G

As with Wi-Fi technology, cellular connectivity is also well known and popular with everyone on Earth. Consumer mobile phones use cellular connectivity to make calls, send **Short Message Service** (**SMS**) messages, and for data services (that is, the internet).

Cellular connectivity provides three basic services to their consumers: voice, SMS, and data. The data service is the most used option in most IoT solutions and use cases; in some special IoT solutions, a voice service might be needed to make emergency calls—for example, if a car was in an accident, a sensor will detect that and will make a voice call to the emergency services and give the location details of where the accident occurred; and the same in the case of a lift accident. SMS can also be used in some IoT solutions as a way to wake up IoT devices to do aa job, so IoT devices are kept in sleep mode but their communication module keeps listening to the mobile network for an SMS. Once an SMS is received, that SMS might trigger a wake-up procedure inside the device to wake up and do aa job, and so on.

There are lots of benefits of using cellular connectivity technology in IoT and in general. Here are a few of them:

- **It uses a licensed spectrum**, so it is more reliable (no interference), and it has high performance and **quality of service** (**QoS**) compared to other unlicensed options.

- **It is everywhere**, so even if you (or the IoT device) go out of the country, there's a roaming agreement between mobile operators across the globe, so you will have connectivity wherever you or your IoT devices go.

- **It supports network coverage** for hundreds of thousands of IoT devices in towns/cities/districts, and so on.

- **Built-in E2E security and high quality of service**.

- The cellular network is **fully managed** and maintained by mobile operators.

The two main disadvantages of cellular connectivity are **the high power** it requires and **the cost** of the service. The high-power challenge is solved by another cellular connectivity option that we will discuss later, but the cost of the cellular service still has disadvantages. The cost is high simply because of the benefits or the advantages we mentioned about cellular connectivity, so it is a trade-off between QoS and cost.

Cellular technologies have been through different evaluation iterations or generations, as detailed next.

First-generation (1G)

This first emerged in Japan in 1979, and then it was rolled out in the US and the **United Kingdom** (**UK**) a couple of years after that. It only allows voice calls to be made; the radio signals used in that generation were analog. As expected with any new technology, there was a lack of reliability, performance, and security in that generation of mobile networks.

2G

2G mobile networks, based on the **Global System for Mobile Communications** **(GSM)** or the **General Packet Radio Service** (**GPRS**), were introduced to overcome the limitations in the first mobile network generation, and they were built based on the digital signaling technology, which increased security and capacity.

2G allowed users to send SMS and **Multimedia Messaging Service** (**MMS**) messages at a data rate of up to 64 Kbps. 2G supported voice and SMS (circuit switch),), but a new version was introduced called 2.5G, which introduced packet switching that supported data services, and for the first time, users were able to send an email and browse the web from their mobile phone. 2.5G offered a data rate of up to 144 Kbps.

Most mobile network operators have shut down 2G networks as they are now quite old, so if there are still legacy IoT solutions that are based on such legacy connectivity technology, a migration project will be required to move it to newer cellular technology.

3G

3G mobile networks fundamentally changed the way mobile phones work. 3G networks utilize the **Universal Mobile Telecommunications System** (**UMTS**) as their core network architecture, which focuses a lot on data mobile services with less focus on voice. It offers a high-speed data rate of up to 14 Mbps, and the consumer or customer experience changed accordingly with such new technology. Mobile phone users started to have features such as video calls, playing online games, watching live and on-demand TV and entertainment services, and some other cool features. It also offers great latency compared to previous generations.

4G

4G mobile networks or 4G LTE offer the same as 3G offers but do everything at a much faster speed, which includes increasing data rates and decreasing latency.

There are different categories of LTE, ranging from 0 to 9, where 0 is the slowest/lowest power and 9 is the highest. CAT 0 set the ground required for the introduction of Cat-M (aka Cat-M1, LTE-M), which is cellular connectivity on top of LTE dedicated mainly for IoT or Machine to Machine (M2M).

If you need to use a connectivity option that requires a low data rate and low power, then you could go with the Cat-M LTE connectivity option. We will cover Cat-M and other low-power cellular connectivity options later in the chapter.

5G

5G is the latest mobile network generation to introduce huge improvements in speed, latency, and bandwidth.. 5G operates on rarely used radio millimeter bands in the30 GHz to 300 GHz 30 GHz to 300 GHz 30 GHz to 300 GHz range; it uses small cells to improve coverage area, and it also uses beamforming technology, which improves latency.

One of the most important features 5G technology provides is extremely low latency, which will help with real-time IoT applications that require extremely low latency.

Another important feature is the data rate, which reaches up to 10 Gbps, so 5G is 10 times faster than 4G.

There are some other modern architectures combined with 5G technology, such as network slicing architecture, which enables mobile operators to offer on-demand, private, and dedicated 5G networks to their customers to align with **service-level agreements (SLAs)**.

You can think about the architecture of the aforementioned cellular connectivity technologies (3G, 4G, and 5G) in the same way as the Wi-Fi setup in your home. How? In mobile networks, you have a RAN, and you have a mobile core network (core). For a RAN, in each city or town, the mobile operator divides that town into cells (this is why it is called cellular) and then assigns and deploys for a group of cells a base station tower. The base station hosts at its highest point the radio equipment of the mobile operator. This base station is similar to a Wi-Fi router in homes, with a difference in scale (that is, the number of devices connected to it) and the coverage (that is, the range of communication).

Devices that contain a cellular communication module (for example, phones and tablets) from the other end will communicate with such cellular base stations. If we are talking about data, the mobile devices will send data as radio waves to the base station at the end. This interface (that is, mobile to base station) is a radio interface or air interface. Then, the base station will convert the radio waves into digital or relay those radio waves to another base station, and at the end, the base station connects to the mobile operator core network (the same as the Wi-Fi router connected to the WAN network (internet) from the other end) that is responsible for handling the routing and control of such traffic.

Now, let's see another **long-range** wireless connectivity option: satellite.

Satellite

Another connectivity option that could be used for long-range communication is satellite communication. This option is not commonly used in IoT solutions, but it is used in very special cases such as when IoT devices are deployed in the middle of the ocean or the middle of nowhere. It is a very expensive option and requires high power.

These types of long-range connectivity options that we have explained so far are power-hungry and offer great bandwidth and latency, which might be needed in some IoT solutions but not in the majority of IoT solutions that require low power and have low data rates. We can't imagine using 5G connectivity in the case of a smart meter, which requires a very low data rate—that is, a few bytes or **kilobytes** (**KB**) every day or every month. As a final point, their connectivity modules/chipsets are expensive.

Now, let's cover a new set of connectivity technologies that target IoT cases that require low power, a low data rate, and long-range coverage. The next technologies follow what is called LPWAN.

There are two categories of LPWAN—licensed LPWAN networks and unlicensed LPWAN networks. Here, licensed refers to the radio spectrum used in the LPWAN network. Let's start with licensed LPWANs.

Cat-M/LTE-M

Classical cellular connectivity options such as 4G and LTE networks consume too much power and don't fit with applications requiring a small amount of data to be transmitted infrequently.

Because of such requirements, a new type of cellular connectivity option emerged to cover such cases. One of those options was Cat-M (aka LTE-M).

Cat-M technology is, in fact, built on top of LTE technology, which is well-known, secure, mature, and future-proof cellular connectivity technology. As explained before, in LTE, there are different categories or releases as per the **3rd Generation Partnership Project** (**3GPP**). Cat-0 set the foundation for Cat-M in terms of reducing cost and power consumption.

The greatest advantage of Cat-M technology is its compatibility with existing LTE networks, so mobile operators don't need to spend too much money on new radio equipment, new site deployment, new core networks, and so on—it's just a matter of a software patch applied to existing LTE networks to support Cat-M.

NB-IoT

NB-IoT is new cellular technology dedicated to IoT, as the name suggests. It is not like Cat-M as NB-IoT uses different radio technologies (**direct-sequence spread spectrum (DSSS)** modulation) from the radio technology used in LTE, which means NB IoT doesn't operate in the LTE band, so operators will have to pay for such new technology in terms of new sites, new radio, and so on to support that new technology.

EC-GSM

Extended Coverage-GSM (EC-GSM) is another option for IoT application use cases but this time in the GSM network, not in LTE.

Now, let's cover some **unlicensed long-range LPWAN options**—in other words, alternatives to the IoT LPWAN cellular connectivity options.

LoRa/LoRaWAN

Remember the wireless or radio concepts we discussed earlier? If I tell you now that you have a free radio spectrum that you can use for launching your wireless service, you will go and build your own wireless network. **LoRa** (or **Long-Range connectivity technology**) built their radio modulation technique, which derives from **Chirp Spread Spectrum (CSS)** technology.

LoRa, as explained earlier, is just for the network physical layer. The **LoRa Alliance** (a nonprofit association with over 500 members globally among **telephone companies (telcos)**, system integrators, and manufacturers) introduced **LoRaWAN**, which is an implementation of the MAC layer that works on top of LoRa. As with any other MAC layer protocol, LoRaWAN focuses on traffic or transmission control—that is, retransmission, message format, transmission error handling, and so on.

LoRa/LoRaWAN network architecture (and similar types of networks, including your own wireless network if you build it yourself) is very simple. It consists of the following:

- **Endpoint devices or nodes**: These are low-power devices that have the LoRa radio communication module that connects to the nearest **LoRa gateway**. LoRa radio chips are only manufactured via the Semtech company, but they are also bundled into modules by many other companies. LoRa modules are much cheaper than cellular modules but a bit expensive compared to Zigbee and Z-Wave modules.

  The LoRa Alliance provides LoRaWAN certification for endpoint devices. Certified endpoint devices provide users with confidence that those devices are reliable and compliant with the LoRaWAN specification.

- **A gateway**: As with the concept of a gateway in short-range technologies such as Zigbee, a LoRa gateway is a device that receives radio signals from LoRa endpoint devices. If you think about cellular networks, the gateway is the base station. The gateway is connected to the **LoRa Network server** through an IP backbone using Ethernet or even cellular (3G/4G/5G) connectivity.

There are two types of LoRa gateway, as outlined here:

- **Indoor gateway**, used to cover and reach difficult-to-reach indoor locations, so that kind of gateway (such as Wi-Fi routers) is usually used for homes or businesses.

- **Outdoor gateway**, which is like a cellular base station. It can be deployed in tall buildings, rooftops, and so on.

- **A network server**: As with a cellular core network, in LoRa, the network server will route the message from LoRa endpoint devices to the right destination—that is, the backend application.

- **A join server**: This is the server that is responsible for storing root keys, generating session keys, and communicating those credentials securely to the network server and application servers. In short, this is the component responsible for the security and authentication of devices in the LoRa/LoRaWAN network architecture.

- **An application server**: This is the server that handles LoRaWAN application layer features such as data decryption and decoding, downlink queuing, MQTT, and HTTP.

Here is a typical LoRa/LoRaWAN architecture:

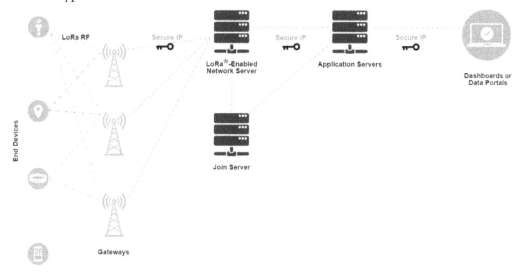

Figure 2.14 – Typical LoRaWAN network implementation

The network topology used in LoRa is a star-on-star topology.

To summarize, there are several benefits of LoRa/LoRaWAN technologies, including the following:

- A **low-power**, **low-cost**, and **long-range** connectivity option (distance of over 10 km in rural areas and up to 3 km in dense urban areas).

- It uses a **free spectrum** and **open standards**.

- **E2E security**: LoRaWAN ensures secure communication between the end device and the application server using **Advanced Encryption Standard 128 (AES-128)** encryption.

- **High capacity**: LoRaWAN network servers handle millions of messages from thousands of gateways.

- **Roaming**: LoRaWAN end devices can perform seamless handovers from one network to another.

Sigfox

Sigfox networks are similar to LoRa networks—that is, both use a free or unlicensed spectrum, the same network architecture, the same network topology, and so on—but if you think about the LoRa network deployment model, it is one you kind of build yourself. You might find some companies build such LoRa/LoRaWAN networks and onboard or sell them to different **small-medium enterprises (SMEs)** to use in their IoT applications and solutions, but in the majority of cases (for example, animal conservation, water conservation, smart farms, and airport tracking), they're typically build-yourself LoRa networks. At the other end, the Sigfox network deployment model is closer to the cellular network deployment model where the mobile operators take care of their cellular network infrastructure, protocols, and so on, and you—as a user of that cellular connectivity—use their network and the different connectivity services. Sigfox has the same concept: the company (Sigfox) runs and manages its LPWAN network, and if you want to use unlicensed and cheaper long-range connectivity, then you can use Sigfox.

Sigfox deployed its network in many countries across the globe and it keeps increasing its network footprint, but clearly, it is not like the footprint of cellular networks.

Summary

In this chapter, you learned about different IoT connectivity options, covering different aspects such as short-range versus long-range, low-power versus high-power, and licensed versus unlicensed radio spectrums. You also learned about the different connectivity protocols and standards that are used in different connectivity technologies. The protocols and standards explained in this chapter are mainly used in the physical and data link layer. IoT application layer protocols such as MQTT and CoAP will be covered in another chapter in this book.

Now, you will be able to identify what kind of connectivity option you should use and leverage for your IoT solution.

In the next chapter, we will continue to look at the anatomy of the word IoT. In this chapter, we covered the letter I. In the next chapter, we will cover the letter T which refers to things—in other words, IoT devices.

3
The "T" in IoT – Devices and Edge

Now, we've come to the last letter of the word "IoT," which is the letter "T", referring to *Things* or *IoT devices*. This is another interesting topic, so let's deep-dive into it.

In IoT solutions, devices can be anything capable of sensing and/or actuating a physical object, have some processing capacity, and are connected to a local area network or the internet. Your smartphone, for example, can be considered an IoT device. Most smartphones have fingerprint (rear-mounted), accelerometer, gyro, proximity, and compass sensors. They also have operating systems such as Android and iOS and have applications that run on top of such operating systems. They are connected devices and can be connected to the internet or a local area network, such as a Wi-Fi network.

In this chapter, we will cover the different topics around IoT devices. The smartphone example mentioned previously was just to explain the concept or the idea. In reality, when it comes to large-scale IoT solutions, IoT devices usually come in the form of **Microcontroller Units (MCUs)** or a single board that contains a microcontroller or a microprocessor and other components. Microcontrollers are minicomputers that have input peripherals, which usually have sensors or actuators connected to them. They also have a microprocessor for processing data that comes through the input peripherals and have output peripherals too, which usually have components such as alarms and LEDs connected. They are then connected to a local area network or the internet for sending and receiving data and messages with an IoT remote control system (such as the IoT backend cloud).

In *Chapter 1, Introduction to the IoT - The Big Picture*, we discussed IoT solution reference architecture, the first two layers of which were **IoT devices** and the **IoT edge**. We will cover these two layers in this chapter.

In this chapter, we will cover the following topics:

- Microcontrollers
- A **Real-Time Operating System (RTOS)**
- The IoT edge

Microcontrollers

Microcontrollers are not new; they have been available for decades now. They are everywhere in everyday consumer products, such as ovens, refrigerators, toasters, mobile devices, televisions, cars, ATMs, printers, and so many other devices around us. They are usually hidden or buried in such products.

In the next few sections, we will cover microcontrollers in very high-level detail without going into low-level electronics or hardware details, which are beyond the scope and goal of this book.

The main goal of this chapter is to make you – as an IoT solution architect or designer – aware, to some extent, of how IoT devices work in large-scale and production-grade IoT solutions at a high level.

Designing and developing an IoT device is out of the scope of this chapter and this book, so we will not discuss things such as **Printed Circuit Board** (**PCB**) design and manufacturing; we will assume that part is typically done by specialized electronic and hardware engineering teams or third parties. In large-scale and production-grade IoT solutions, tasks such as IoT device manufacturing and connectivity are usually outsourced to external specialized suppliers and third parties.

Even at a software level, for a long time, we used to have dedicated and specialized software developers with special skills focusing on designing and building embedded software. Embedded software developers used low-level programming languages to run embedded software in embedded systems and microcontrollers. Other software developers specialized in building enterprise application software, such as web and mobile apps that run on browsers, mobiles, or general-purpose PCs using high-level programming languages.

Now, you might not need such specialization in the device software domain because enterprise developers can now also write software running through microcontrollers. This is possible due to the rich availability of modern **Integrated Development Environments** (**IDEs**), supportability and continuous enhancements, and new features of high-level programming languages such as Python, Java, C++, and C#.

Now, let's understand what a microcontroller is and how it fits into IoT solutions.

What is a microcontroller?

In short, a microcontroller is a miniature or very small computer that includes – in general – a microprocessor (CPU), memory, and input/output peripherals, all integrated into a single chip. This chip is usually called an **Integrated Circuit** (**IC**). In other words, if you compare traditional PCs or server hardware with a microcontroller, you will find that both use the same kind of components (a CPU, memory, and so on), but for traditional PCs/servers, those components come as separate units or chips, while in the microcontroller, all those components are integrated into a single chip.

The microcontroller chip can be embedded with other devices due to its small size. The typical dimensions for MCUs are a height ranging from 0.5 mm up to 4.95 mm, and a length ranging from 4 mm up to 35.56 mm, so it is a very small computer.

There are many different packaging types for microcontrollers, such as a **Dual In-Line Package** (**DIP or DIL**), which is an electronic component package with a rectangular housing and two parallel rows of electrical connecting pins, a **Quad Flat Package** (**QFP**), a **Ball Grid Array** (**BGA**), a **Chip Scale Package** (**CSP**), and other packaging types.

Some people get confused between **microcontrollers** and **microprocessors**. This might be because of the common word micro used in both terms and also the same meaning of the words processor and controller. As I explained previously, a microcontroller contains a microprocessor, which is just one of the components in the microcontroller single-chip, while the microprocessor is used with general-purpose computing devices in addition to a microcontroller chip, and its job is very clear in all cases, as we will explain later.

There are many microcontroller chips, all having different sizes, costs, architecture, packaging types, and supported features. For example, the ESP-WROOM-32 microcontroller in the following figure (`File:Espressif ESP-WROOM-32 Wi-Fi & Bluetooth Module.jpg`) by Brian Krent, licensed under CC BY-SA 4.0CC BY-SA 4.0) is considered to be a powerful, rich, and generic microcontroller with built-in support for Wi-Fi, Bluetooth, and **Bluetooth Low Energy** (**BLE**) connectivity options. It is commonly used in a wide variety of IoT solutions; lots of development kits on the market for prototyping and proof of concept purposes come built-in with this microcontroller:

Figure 3.1 – ESP-WROOM-32

The main microcontroller components are as follows:

- **The microprocessor or the CPU (Central Processing Unit)**: This is the brain of the microcontroller. The microprocessor performs the basic arithmetic, logic, I/O operations, and communication with all other components in the microcontroller, similar to what it does in general-purpose computers.

  In the preceding photo, the ESP-WROOM-32 microcontroller has a Ten silica Xtensa LX6 microprocessor.

- **The memory**: The microcontroller has two main categories of memory, just as in general-purpose computers. The first one is **Read-Only Memory** (**ROM**), which is used in storing the application programs. It is pre-written permanent memory that persists even if the power is lost. The second type is **Random Access Memory** (**RAM**), which holds data during microprocessor work for a short duration, so it's a kind of temporary storage used only when the power is on, and the data will be lost if the device is turned off or the power is lost.

  In the preceding photo, the ESP32-DevKitC microcontroller has a 4 MB flash and a 520 KB SRAM memory. Flash memory is a type of **Electronically Erasable Programmable Read-Only Memory** (**EEPROM**), which is a mix of ROM and RAM. The "S" in SRAM stands for Static, as there are two types of RAM, static and dynamic.

- **The input/output peripherals or pins or ports**: A microcontroller has many I/O peripherals. They allow the microcontroller to be linked to other components and circuits to handle the flow of input/output data signals and a power supply. Sensors and actuators are usually attached to these I/O peripherals – for example, you can attach an ultrasonic sensor to one of the input peripherals and start reading and processing this sensor reading. You can also generate output messages or alarms based on that input. These outputs can be in the form of switching on an LED component that is attached to one of the microcontroller output pins or turning on a servo motor component attached to one of the microcontroller output pins, and so on.

  There are different numbers of pins in microcontrollers. For example, some microcontrollers offer 8 pins, while others offer 16, 32, and up to 100 pins, and this number might even increase in the future. In each data sheet of a microcontroller, there is what is called a pin layout. It explains the number of pins in the microcontroller in detail and the functionality of each pin or group of pins.

There are some other optional components that can be attached to a microcontroller to add more capabilities or features. They are as follows:

- **ADC (Analog to Digital Converter)**: As the name suggests, this component converts analog signals to digital. For example, in the ESP32-DevKitC microcontroller, there are some pins called analog pins, which you can connect an analog sensor to, such as an ultrasonic or temperature sensor. The values generated from such sensors are analog, and then the ADC will convert such analog values to digital values for the microprocessor to process.

- **DAC (Digital to Analog Converter)**: This component performs the opposite function to the preceding ADC component. In the ESP32-DevKitC microcontroller, there are some pins called digital pins used for that purpose.

- **System buses (data and address)**: The different components of the microcontroller talk to each other through these buses.

- **Timers/counters**: These handle the timing, counting functions, modulations, and so on.

- **Interrupt controller**: This provides an interrupt for a working program; it can be an internal interrupt or an external interrupt.

- **Micro-USBs**: These are used usually to power a microcontroller, as in the case of the ESP32-DevKitC microcontroller. They are also used for uploading software binaries to the microcontroller.

- **Serial ports**: These are used for receiving and transmitting serial data.

There are so many other components that we can't list down here, as they differ from one microcontroller to another. You can usually get such detailed information by checking the data sheet of the microcontroller.

In *Chapter 2, The "I" in IoT – IoT Connectivity*, we talked about IoT connectivity and also about IoT communication modules in IoT devices. The communication module is one of those add-ons or optional components of microcontroller architecture. For example, in the case of the ESP32-DevKitC microcontroller mentioned earlier, it supports Wi-Fi (802.11 b/g/n) and BLE connectivity options.

On the market today, there are so many microcontroller manufacturers, brands, types, and architectures. In the next section, let's look at the different microcontroller types available.

Microcontroller types

Microcontrollers can be classified based on their architecture type or the number of bits they can support at any time. For example, we have the following microcontroller types based on bits:

- **8 bits**: This is a very basic and limited microcontroller and usually very cheap. It supports basic logic and arithmetic operations.

- **16 bits**: In comparison with the 8-bit microcontroller, the 16-bit offers much better performance and accuracy.

- **32 bits**: This is much better in terms of performance, efficiency, and accuracy.

There are two architecture types for the microcontroller, the **Harvard Architecture**, and the **von-Neumann architecture**. Without going into low-level detail about those two architectures, the difference between them is basically how data is exchanged between the CPU and the memory.

As stated earlier, microcontrollers and electronics are topics that cannot be discussed in detail here, but let's focus in the next section on the main criteria you should use for selecting IoT microcontrollers or IoT devices for production-grade and large-scale IoT solutions.

IoT microcontroller selection criteria

There are so many business and technical factors for selecting the best-fit IoT microcontroller for your IoT devices. Here are just a few:

- **Power consumption**: We discussed in *Chapter 2, The "I" in IoT – IoT Connectivity*, the kind of power needed for your IoT devices. Do you need a low-power (battery-based) device or do you expect to design an IoT product that has to be plugged into the main power supply all the time so that you don't have to worry about power consumption? If you are looking for a low-power device, then you might go for an 8-bit microcontroller. If so, this type of microcontroller might reduce power consumption, but at the same time, it has a limitation in data storage and processing, so you need to make a trade-off decision here between power consumption and other features such as processing and storage.

- **Cost**: In the smart parking use case, if we need to bury under each parking lot a sensing IoT microcontroller, then we have to multiply the number of parking lots we have by the number of parking areas in a city to determine how many microcontrollers we need for our solution. In this case, you may find that a lot of microcontrollers are needed, so you will have to choose. A very cheap microcontroller that can do the simple function of sensing the presence of a car in the parking lot and sending such information to the IoT gateway device (or the IoT cloud, if applicable) can be used. The IoT gateway device itself might use an expensive microcontroller, which should be fine, as typically, gateway devices needed in an IoT solution are fewer compared to endpoint IoT devices, so the cost will not be a problem.

- **Security**: This is a very important point. In production-grade IoT solutions, you can't just purchase some dummy and cheap microcontrollers and deploy them into a production line, as this would be very dangerous. Such dummy or basic devices are usually easily hacked. In a production-grade solution, IoT devices and microcontrollers must pass many security tests and be certified by many security regulations and standard bodies to confirm that they are secure.

As mentioned earlier, microcontrollers are embedded in multiple consumer devices. In other words, they might be in the hands of an end user, so you should make sure security controls such as avoiding reverse engineering, protecting your company's **Intellectual Property Rights (IPR)**, as well as memory protection, tamper protection, and many other microcontroller security aspects (at a hardware level) are in place before distributing and deploying IoT devices in the field.

- **Connectivity support**: You need to check whether an IoT microcontroller comes with the built-in connectivity option that you are looking for, or whether you have to purchase the communication module you require separately and attach it to your microcontroller.

- **Size**: You also need to consider what your size requirements for the microcontroller are and where it will be embedded.

- **Performance**: What are the performance requirements for the microcontroller? What is the expected processing time? If you need more processing, then you should be aware of other requirements such as low power, as the more processing you need, the more power will be consumed.

- **Future-proof and flexibility**: Things such as the number of input and output pins in a microcontroller give flexibility in design, and the more pins or ports you have for input and output, the more you can customize the microcontroller to fit with future requirements – for example, you might need to add a sensor to the device. So if there's no port for that, you will have to change the whole microcontroller.

- **Harsh physical environmental conditions tolerance**: This is to what level a microcontroller can survive in severe environmental conditions such as very high temperature, high radiation, and similar severe environmental conditions.

- **Brand**: Working with a well-known microcontroller brand or manufacturer will ensure that features such as support, security, maintenance, and quality are reflected in the microcontroller board or chip that you get from a big-brand name.

There are many microcontroller brands or manufacturers on the market today, such as **Espressif Systems** (the ESP32-DevKitC development board used as an example in previous sections is from this brand/company), the **Epson Semiconductor, Microchip, NXP**, and **Texas Instruments** and so many other available brands.

There are also many **development kits** available on the market, which are cheap and usually used for prototyping and **Proof of Concept (POC)**. The most famous one is **Arduino**.

Now, we are done with the **hardware part** of a typical IoT device, which, in short, consists of a microcontroller chip, sensors, and actuators attached to the microcontroller chip through its I/O peripherals.

The question now is, what about the **software part** of that IoT device, which is responsible for processing the sensor, reading data, and sending it to the edge or IoT cloud, or the software that reacts to the instructions or jobs that are sent to the IoT device from the IoT cloud?

In the next section, we will answer that question by explaining an RTOS and its benefits for IoT device software development.

An RTOS

Before we talk about an RTOS, let's talk briefly about embedded software in the next section.

Embedded software

Enterprise or traditional software applications are usually written in a high-level programming language (human-readable languages) such as C, C++, Java, and C#. Then, these high-level programming languages are converted into a low-level machine-readable language, or in short, they are converted into a series of 0s and 1s at the end. Computers understand only binary data (0s and 1s).

When it comes to embedded software that runs on a microcontroller, it is the same. You need IoT device application code to be converted into 0s and 1s, and then the microcontroller or the microprocessor of the microcontroller will load that program into the memory of the microcontroller and start executing it.

If you have played around with the Arduino microcontroller before, I am sure you realize how it is easy to write software that runs on it and interacts with different Arduino board peripherals (such as sensors and LEDs). If not, no problem. The following steps are the ones typically used in writing and running embedded software on Arduino at a high level:

1. You have your personal development workstation, which can be based on Windows, Linux, or macOS, on the one hand, and you have the Arduino microcontroller board, on the other.

2. Connect your development workstation with an Arduino board to the USB cable, using the USB peripheral that came built-in with the Arduino board.

3. In your development workstation, you need to download and install the Arduino IDE, which, similar to Eclipse, Microsoft Visual Studio Code, and other IDEs, you can use to `write`, `build`, `debug`, and `run` software code in a programming language of your choice.

4. In the Arduino IDE, you write an Arduino **sketch**, which is a kind of code skeleton or code template showing where you should write your code logic. In the sketch, you have two mandatory methods. One is `Setup ()`, which runs only once upon the boot of the Arduino board and is usually used for initializing the state of the Arduino board – for example, initializing and setting which Arduino microcontroller pin will be used for input or output, and initializing variables. The other method is called `Loop()`; this is the main method, where your code logic is set and run. As the name suggests, your code logic will run in a loop forever till the microcontroller power is turned off or lost.

5. Once you have written the code (in C++) in the Arduino sketch, using the Arduino IDE, you can verify whether the code is correct. If it is, upload it to the Arduino microcontroller board. This step is done in the Arduino IDE by selecting the microcontroller board that it is connected to.

6. Behind the scenes, what the Arduino IDE is doing is simply converting the high-level code that you wrote into machine code (0s and 1s), which can be run by the microcontroller microprocessor.

From the preceding steps, we see now how easy it is for enterprise software developers to write embedded software and interact with microcontroller peripherals and components. Previously, embedded software required special skills in coding to deal with limited and constrained microcontrollers or embedded systems.

You can use the Arduino IDE with other boards aside from Arduino – for example, the ESP32-DevKitC microcontroller mentioned earlier is supported by the Arduino IDE.

You need to check the microcontroller data sheet to see how you can do development on top of it. Some microcontrollers provide an IDE and different development toolkits for simplifying software development on their microcontrollers.

But the question is – can we have software running on hardware directly instead of running on top of an operating system? The answer is yes, as we explained earlier and in the Arduino steps mentioned previously. Software is run directly into the microcontroller. If you think about it, an operating system itself is a kind of software that runs directly on hardware and interacts with different hardware components, such as a processor and RAM.

In the next section, we will see how the operating system helps in accelerating embedded software development and adding lots of benefits to it.

Operating systems (traditional versus real-time)

Operating systems, in general, provide an abstraction layer between the different applications run on top of them and the underlying computer hardware.

Such abstraction or resource management is one of the greatest benefits that operating systems provide as you don't need – as an application coder – to write every time, aside from your business or core software application, another set of hardware drivers to make your application work. With an operating system, you simply focus on writing the business or core software applications, and the operating system takes care of the hardware management part.

You might be wondering why we need an operating system for such low-value and constrained IoT devices. To answer that, we need to refer back to earlier when we explained how to write embedded software that runs directly into a microcontroller board. With that type of software, you need to handle the connectivity software stack, security, middleware platforms such as filesystems, resource management, memory management, event handling, power management, messaging, interrupt handling, task management, and the portability of the software when you need to move it to another microcontroller board. So this approach (writing software directly into a microcontroller board) is clearly not the right one for production-grade, large-scale IoT devices. It can, however, be used for development or prototyping purposes, but for production-grade devices, you will need the cross-cutting or non-business services mentioned previously to be abstracted, managed, and provided out of the box by another software platform, which in this case will be the operating system.

So, it is clear now that we need an operating system for the IoT device, but the next question is, *can't we just use or install traditional and general-purpose operating systems such as Windows and Linux in those IoT devices or microcontrollers?* The answer is, unfortunately, no. IoT devices are very limited and constrained devices in terms of computing resources (that is, memory, processing, and so on), so they need a special type of operating system to handle or deal with such challenges or limitations in computing resources. For example, in the IoT microcontroller world, you might have a processor with only one core. For such a processor, only one thread can be run at a time, so the kernel or operating system should handle and manage that limitation efficiently. This limitation and others are why there is a need for a new type of operating system for such devices, which is called an RTOS.

It is called real-time because it provides a high deterministic real-time response for tasks or threads. In other words, an RTOS does one task at a time and does it well, while in a traditional OS, there is no guarantee when the task will be completed or whether there'll be a response. Such real-time response behavior is required for the majority of IoT use cases – for example, if car sensors detect a collision is about to happen, then a driver's airbag has to be opened immediately without any delay. In traditional operating systems, there's no guarantee that the application you run will respond immediately; this is one of the differences between the traditional operating system and RTOSs.

The main components of RTOS systems are as follows:

- **Scheduler**: This is the component responsible for identifying tasks that should be run.

- **Memory management**: To manage both the kernel memory allocation and the application memory management.

- **Inter-task coordination**: To enable task communication using components such as queues, semaphores, mutexes, and buffers.

In an RTOS, the application designer or coder can set the priority of the task or the thread. Then, the RTOS kernel (the scheduler) will decide which task or thread is to be run based on that priority. You might set the priority high for some tasks, such as the driver's airbag example mentioned earlier (this is known as **hard real-time**), while you can set the priority low for another task that is not urgent or poses no danger to respond slowly (this is called **soft real-time**).

There are so many RTOS (open source and proprietary) systems available on the market today, such as **Contiki,FreeRTOS, MbedOSMicroPythonRIOTTinyOSWindows 10 IoT MbedOS, MicroPython, RIOT, TinyOS, Windows 10 IoT, Apache Mynewt, Zephyr OS, Ubuntu Core 16 (Snappy), Nano RK, LiteOS , OpenWrt, Android Things, MicroEJ OS, Express Logic ThreadX, Mentor Graphics Nucleus RTOS,** and **Particle**.

In the next section, we will go through the RTOS selection criteria, from which we will choose an RTOS system for our IoT device.

RTOS selection criteria

The following are criteria that you should consider when selecting the best RTOS platform for your IoT device:

- **Portability**: What kind of hardware can the RTOS candidate support or run-on top of it? The hardware will usually be different most of the time.

- **Reliability and resiliency**: How reliable is the RTOS candidate? This includes the time it takes to respond to mission-critical events and how resilient it is to errors and system failure.

- **Security**: How many security controls are provided by the RTOS candidate? Does the RTOS candidate support security features such as SSL, secure boot, and encryption?

- **Size or footprint**: How many device resources will the RTOS candidate take? An RTOS system will run on very limited and constrained devices/microcontrollers, so the RTOS candidate should take a very small footprint of such limited resources. A typical RTOS kernel binary image is in the range of 4,000 to 9,000 bytes.

- **Cost**: What is the cost of the RTOS candidate? Is it open source or licensed?

- **Support**: Is the RTOS candidate supported by large tech communities, companies, and third parties? Are well-written documentation and tools available?

- **Modularity or flexibility**: How easy is it to extend the RTOS candidate system or add new plugins to it? In other words, what level of architecture modularity does the RTOS have? An RTOS has a core engine or a kernel, but on top of that, can you plug new modules into the RTOS to support new features that did not come built into the core? This is a very important criterion that you need to check before selecting an RTOS system.

- **Connectivity**: What kind of connectivity software stack or protocols are supported by the RTOS candidate? Connectivity examples include Ethernet, Wi-Fi, BLE, and so on, which we discussed in *Chapter 2, The "I" in IoT – IoT Connectivity.*

- **Development tools and kits**: How easy is it to quickly start development on top of the RTOS candidate system? What kind of development tools are available and supported by the RTOS candidate?

In *Chapter 1, Introduction to the IoT – The Big Picture*, we mentioned in the book's strategy section some of the principles that the book will follow; in the next few sections, we will follow two of those principles. The first principle we mentioned earlier was that we cannot, by any means, cover all available IoT systems or platforms in this book (or in other books), as it is not realistic. The second principle was to use one, and only one, platform throughout the book to explain the concepts discussed, and we chose the AWS IoT platform for this job (only for the sake of simplicity; other IoT platforms such as the Azure IoT and Google Cloud IoT also provide the same IoT services and solutions with the same level of quality). Based on those two principles, we will choose **FreeRTOS** from the previously mentioned RTOS systems and discuss it in the next sections.

I hear you ask how FreeRTOS is related to the AWS IoT platform. The answer is, AWS selects and extends the open source FreeRTOS platform and offers Amazon FreeRTOS as of the AWS IoT services provided to its customers. We might use the terms Amazon FreeRTOS or AWS FreeRTOS in the following sections, but both are interchangeable.

In the next sections, let's discuss first the generic FreeRTOS and then Amazon FreeRTOS.

FreeRTOS

According to the official FreeRTOS website (`https://www.freertos.org/`), FreeRTOS is an RTOS for microcontrollers developed in partnership with the world's leading chip companies over a 15-year period. Now downloaded every 170 seconds, FreeRTOS is a market-leading RTOS for microcontrollers and small microprocessors. Distributed freely under the MIT open source license, FreeRTOS includes a kernel and a growing set of IoT libraries, suitable for use across all industry sectors. FreeRTOS is built with an emphasis on reliability and ease of use.

FreeRTOS is free and open source despite the fact that Amazon has the stewardship of it. This is because Amazon is investing in the development of the FreeRTOS kernel. When we talk later about Amazon FreeRTOS, you will see that it is not like Amazon or AWS fork, where open source projects come with their version of what is called Amazon FreeRTOS. Instead, Amazon supports the development of the main FreeRTOS kernel and also adds different software libraries to the FreeRTOS kernel to securely connect FreeRTOS-based IoT devices to the AWS cloud or AWS IoT Greengrass (the edge). We will explain this later in this chapter.

So, if you are not using the AWS cloud as your IoT cloud, then there's no issue, as you can use FreeRTOS-based IoT microcontrollers or devices and connect them to whatever IoT cloud you have. In other words, you don't need to have the IoT cloud running and hosted in AWS to use FreeRTOS, but if you have the IoT cloud host on AWS, then you will get many benefits from using FreeRTOS, as the software libraries and integration with the AWS IoT cloud are already builtin with Amazon FreeRTOS.

Now, let's talk about Amazon FreeRTOS.

Amazon FreeRTOS

Amazon FreeRTOS is based on the FreeRTOS kernel with some additional or extended Amazon software libraries to integrate with Amazon cloud and Amazon edge services out of the box. This way, AWS helps embedded software developers integrate a device's embedded software smoothly with AWS cloud services, and embedded software developers don't need to be AWS cloud experts to be able to integrate with AWS cloud services, especially AWS IoT cloud services.

The following diagram shows the architecture of Amazon FreeRTOS:

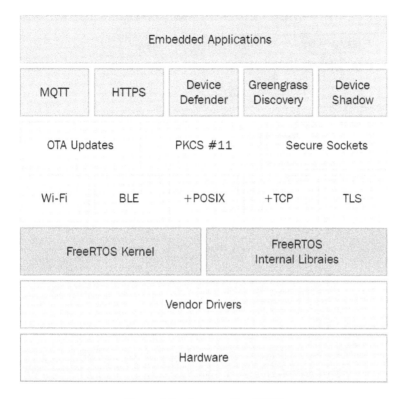

Figure 3.2 – Amazon FreeRTOS

Amazon FreeRTOS architecture is very simple and modular, as shown in *Figure 3.2*. At the bottom, you have the hardware part of the microcontrollers or IoT devices, and then you have the FreeRTOS kernel and its internal libraries. On top of that layer, you have the FreeRTOS porting libraries (WiFi, BLE, TCP, OTA updates, and so on). Those libraries are platform-dependent, and their contents change depending on the selected hardware platform. Then comes the Amazon FreeRTOS application libraries (MQTT, HTTPS, Device Defender, Greengrass Discovery, and Device Shadow). Those application libraries enable integration out of the box with the AWS cloud. Finally, there are the embedded applications, which hold the business logic of your IoT solution and use Amazon application libraries and SDKs for integrating with the AWS cloud.

We will discuss those application libraries in more detail later in the book.

FreeRTOS or Amazon FreeRTOS is packaged as a binary image, which you can download to your microcontroller device.

Amazon FreeRTOS supports the following microcontroller devices:

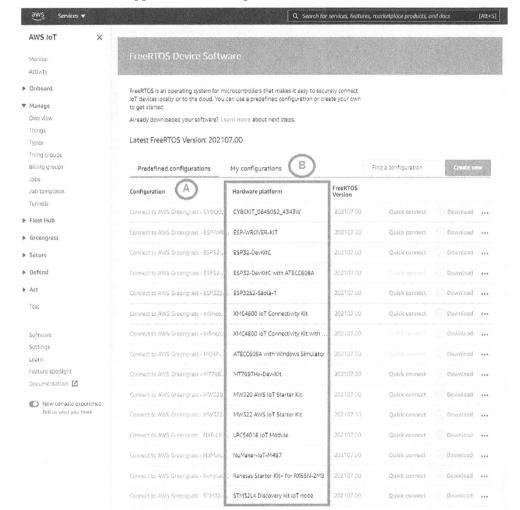

Figure 3.3 – Amazon FreeRTOS software from the AWS Management Console

In the preceding screenshot (*Figure 3.3*), note the following:

- The screenshot was captured from our AWS account. To reach this page in your AWS account, go to the AWS Console and then search for `AWS IoT service`. Then, choose the **Software** link from the left menu and then FreeRTOS Device Software – configure download.

- The list in the picture is not complete; we have cut it for simplicity purposes. There are other hardware platforms supported by Amazon FreeRTOS; you can check the AWS FreeRTOS service in the AWS Console for the complete and latest list.

- The circle with the letter **A** in *Figure 3.3* shows the predefined configuration or built-in software libraries that come with this version of Amazon FreeRTOS, which are supported by the selected hardware or microcontroller device. If you click on this link, you will see those built-in libraries. You can remove some of these libraries if they are optional or removable, and you can also add libraries if needed; we will show that in the screenshot in *Figure 3.4*.

The circle with the letter **B** in *Figure 3.3* shows you that you can customize or build your own version of Amazon FreeRTOS by selecting the libraries you want:

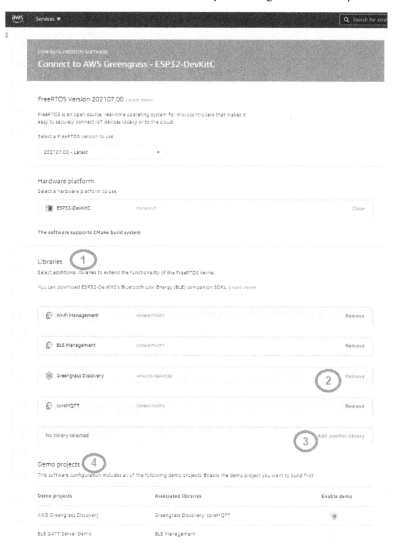

Figure 3.4 – Amazon FreeRTOS software – configuring libraries

In *Figure 3.4*, note the following:

- The circle with the number **1** shows a list of libraries that come pre-defined or built-in with that version of Amazon FreeRTOS.

- The circles with the numbers **1** and **3** show the ability to customize those libraries by removing and adding additional libraries.

- The circle with the number **4** shows the list of **Demo projects** that come pre-defined with that Amazon FreeRTOS version. You need to select one of these demo projects to add to the Amazon FreeRTOS image, which you can download and install on the IoT device. This is a good option, as it enables you to start developing Amazon FreeRTOS quickly.

Now, let's recap the selection and setup of an IoT (an endpoint or leaf) device in the IoT solution:

- **Sensors**: Based on business requirements, you will be able to list down what kinds of sensors (for example, temperature, humidity, pressure, proximity, and accelerometers) you need to collect the data you are looking for from the real world. This is called sensor selection criteria. Questions such as *which temperature sensor is the best one* are beyond the scope of this book. Some microcontroller boards come with some built-in sensors, which might be enough for what you are looking for. See the next point about microcontrollers.

- **Microcontrollers**: In the previous sections, we went through microcontrollers and the criteria you should follow while selecting a microcontroller board for your IoT device.

- **An RTOS and embedded applications or software**: In the previous sections, we went through the benefits of having an RTOS flashed and installed on a microcontroller; then, the embedded applications can run on top of that RTOS. We also went through the different criteria that can be used in selecting the best RTOS for an IoT microcontroller device.

In the next sections, we will go through the IoT Edge, covering its location, hardware, and software requirements.

IoT edge

In *Chapter 1, Introduction to the IoT – The Big Picture*, we briefly talked about the IoT edge and why we need an IoT edge cloud. In this chapter, we will dive deeper into the IoT edge cloud in terms of IoT edge location, hardware, and software.

IoT edge – location

The location where edge devices will be located is very important. The closer an edge device is to IoT endpoint devices or the source of the data, the more benefits you will get, such as very low latency and responsiveness.

The connectivity options used in an IoT solution play a critical role in choosing an edge location, as explained in *Chapter 2, The "I" in IoT - IoT Connectivity*. The communication range of the connectivity option used is what determines the location of the edge device – for example, some of the connectivity options we discussed in *Chapter 2, The "I" in IoT - IoT Connectivity* have a very limited range. In this case, the edge has to be very close to the IoT endpoint devices and so on.

Also, IoT edge hardware plays a role in identifying the location of an edge device. The smaller the edge device, the closer you can place it next to IoT endpoint devices. If it is large, then you need a dedicated room or facility to host it.

We have many options for edge location, such as the following:

- Within the same area as the deployed IoT endpoint devices – for example, in a smart store, you can have the edge device sitting in one room in the same store. The edge device in the store example could be centralized, meaning that one edge device serves the whole store, or could be distributed on multiple floors in the store – for example, for the first floor, you have the first edge device serving the IoT endpoint devices deployed on the first floor.

- You might have a big or medium edge facility that can't be hosted within the same space where the IoT endpoint devices are deployed, so the option with the support of long-range connectivity technologies is to deploy that edge facility a bit further from the IoT endpoint device, which might be within the same district or the same city.

  In that option, AWS, for example, has a service called AWS Wavelength, which, from an AWS customer point of view, is a zone for deploying AWS resources, but behind the scenes, AWS partners with mobile operators host that service or zone at mobile operator edge premises. Mobile operators have many edge locations, which are naturally driven by the nature of cellular connectivity technology where a whole city is covered by a mobile operator infrastructure.

- The last option of a location for the edge might be within the same country (for example, within England). We can have the option of within the same geography region (for example, within Europe), but this will not help with IoT edge requirements (low latency, data locality, and so on).

In the next section , let's discuss IoT edge hardware requirements.

IoT edge – hardware

The hardware of the IoT edge can be as small as a microcontroller or a small-scale data center. It depends on the workload you intend to run on the edge. A workload might involve a small amount of data processing that comes from IoT endpoint devices before sending it or relaying it to the IoT backend cloud. For such a workload, a microcontroller or minicomputer is suitable. Some workloads might require processing, real-time analytics, machine learning at the edge, low-latency apps, and so on. For these kinds of workloads, you will require quite strong hardware to run those services without any compromise on performance, resilience, and security.

In the next section, let's discuss IoT edge software.

IoT edge – software

Let's think first in a generic way about the common components and solutions that are required in the IoT edge cloud. In doing that, we have a few options and services.

Generic solutions and services

The following are a few of the generic core solutions and services that are typically required in the IoT edge:

- **IoT gateway**: You need an IoT gateway, whether it is a hardware-based or software-based one. The IoT gateway will cover the following gateway features:

  - **Security**: Securing the connectivity between the IoT endpoint devices and the IoT edge and IoT backend clouds implicitly. In other words, when you secure the touchpoint or the front zone, you implicitly also secure the backend zone.

  - **Protocol transformation**: Transforming legacy protocols into new modern and supported protocols, offered and supported by the IoT edge cloud or the IoT backend cloud. This is a very important feature, and you will need it in the modernization or transformation of brown-field IoT projects – which might have legacy IoT devices that are using legacy protocols and technologies – to new technologies.

  - **Data enrichment and preprocessing**: To enrich data that comes from IoT devices before sending it to the IoT edge cloud or the IoT backend cloud for further processing. This feature can help also in saving bandwidth and money by discarding worthless business data at the edge.

  - **Other standard gateway features**: This includes routing and filtering.

- **Message broker**: At the edge, the endpoint IoT devices can talk to each other or the IoT edge or the IoT backend cloud directly. We can put the IoT backend cloud aside for now, since we are talking here about the IoT edge cloud (we will discuss the IoT backend cloud anyway in another chapter in the book). You will need a message broker with a **publish-subscribe (pub-sub)** paradigm deployed on the IoT edge cloud to receive the messages from the IoT endpoint devices and route them to the other IoT endpoint devices in the local network without a need for internet access to the IoT backend cloud (this can be classified as offline mode of the IoT solution).

  MQTT is one of the best IoT application protocols, which fits well with the common IoT low-power and very constrained devices, so what we are talking about here is an MQTT message broker to be deployed on the IoT edge.

  You might have a case where your IoT endpoint devices use a different protocol to MQTT, such as the **Open Platform Communications Unified Architecture (OPC-UA)**, which is an information exchange standard for **industrial communication**. What you should do in this case is simply use the IoT gateway – as explained earlier – to translate whatever protocol IoT endpoint devices use into the MQTT protocol, which is supported (or should be supported) by the IoT edge cloud and also the IoT backend cloud.

- **Compute or local processing**: You might need to run some business logic at the edge. Business logic is usually captured in the form of a software application that requires a hosting environment or a runtime environment to run on top of it. ML inference capabilities come under this compute or processing category. The ML model is a software program that has been defined and trained in the cloud where you have massive compute and storage resources, while on the edge, you just need to run that ML program or model.

  The easiest thing here is to have a dedicated server or VM running at the edge, but the edge, in general, is not constrained like a data center, as we explained in the IoT edge hardware section. You might have all your edge workload run on a small device such as a Raspberry Pi due to your edge device needing to be used outdoors, for example. Even if it is indoors, you still have a very limited space to host that edge server; hence, containers are usually the best solution at the edge when it comes to hosting and running software applications.

  Containers have a small footprint compared to VMs, and with a strong container orchestration platform such as Kubernetes, you can run different software applications at the edge smoothly and efficiently. In fact, containers pave the way for serverless architecture at the edge; we will see that next when we talk about AWS Greengrass.

- **Storage**: You don't need a massive storage service at the edge, but you do need temporary storage to store the IoT data generated from IoT endpoint devices for a short period of time.

  There are different reasons why you need to store the data at the edge for a smalltime window; the main reasons are as follows:

  - **Connectivity**: The main reason for, and objective of, the IoT edge is to keep IoT devices connected and doing their work, even in the case of losing internet connectivity. So, the IoT edge will buffer a message till internet connectivity returns, and then it will send (or sync) that buffer of the data to the IoT backend cloud.

  - **Real-time analytics**: For real-time analytics, you need a small window of data. You might have queries run on the last 10 minutes of data and so on. In other words, you don't need traditional batch analytics where you wait until the end of the day or week to run business analytics queries.

  - **Security**: You need to store some security credentials to cover local authentication and authorization scenarios when losing internet connectivity with the IoT backend cloud. You will need a local cache of the required security tokens and credentials to allow the flow to continue when connectivity is back, and you can then sync with the IoT cloud, the state, and so on.

- **Rule Engine and Complex Event Processing (CPE)**: This component is responsible for processing and orchestrating events or data streams that come from IoT endpoint devices. Complex rules might be applied for such incoming IoT data streams.

- **Stream processing engine**: This engine is required as part of a real-time analytics solution at the edge.

The previously mentioned components define the generic solutions and services required at the edge.

Now, in the next section, we will explain the different options for implementing IoT edge software solutions and services.

Implementation options

There are many ways to implement an IoT edge software solution, such as the following:

- **Using a Do-It-Yourself (DIY) model**: This is where you implement the whole edge software solutions and services in-house or in partnership with a third-party vendor. The solution itself might use open source-only components or **Commercial Off-the-Shelf (COTS)** products or have a mix of both types of products.

- **Using public cloud edge offering solutions**: If your IoT backend is hosted or managed by IoT public cloud services such as AWS, Azure, and Google, then most of those public cloud providers offer an edge solution that covers all the software solutions and services that are required at the edge, and also offer smooth integration between their edge solution and their public cloud services. Amazon provides AWS Greengrass, while Microsoft provides Azure IoT Edge.

In the next sections, we will talk about the AWS Greengrass service.

AWS Greengrass

According to the AWS official documentation, AWS IoT Greengrass is an open source IoT edge runtime and cloud service that helps you build, deploy, and manage IoT applications on your devices. You can use AWS IoT Greengrass to build software that enables your devices to act locally on the data that they generate, run predictions based on ML models, and filter and aggregate device data. AWS IoT Greengrass enables your devices to collect and analyze data closer to where that data is generated, react autonomously to local events, and communicate securely with other devices on a local network. Greengrass devices can also communicate securely with AWS IoT Core and export IoT data to the AWS cloud. You can use AWS IoT Greengrass to build edge applications using pre-built software modules, called components, which can connect your edge devices to AWS services or third-party services. You can also use AWS IoT Greengrass to package and run your software using Lambda functions, Docker containers, native operating system processes, or custom runtimes of your choice.

An AWS IoT platform that runs on the AWS cloud is one of the best IoT platforms on the market. We will explain IoT platforms in a dedicated chapter later in the book. What AWS did regarding edge solutions was simply push users to run the IoT services and solutions that they run on the large-scale AWS IoT cloud **locally** on an edge device. Those services are packaged as a software binary that is installed on the supported edge device.

This is the architecture of AWS IoT Greengrass:

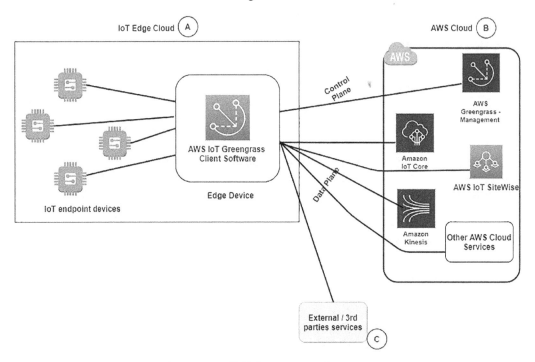

Figure 3.5 – AWS Greengrass architecture

The architecture in the preceding diagram has the following:

- The circle with the letter **A** shows the IoT edge side or the ground/field where the IoT endpoints and edge devices are deployed. The edge device will have the AWS Greengrass software installed, configured, and running.

- The circle with the letter **B** shows the AWS cloud side, which is divided into two main parts:

  - One part shows the **management part** of the AWS Greengrass service, which runs on the AWS cloud, not on the edge. In that management service, you can configure and manage the AWS IoT Greengrass client software, which is running or will be running on the edge device. You can configure and deploy the different components, which will be run locally on the AWS IoT Greengrass client software/ runtime, as we will explain later. You can configure and deploy an ML model on the edge, and update the Greengrass software and so many other Greengrass client software management features that are handled and managed from the AWS IoT Greengrass cloud service. This part is the **control and management plane** of the AWS IoT Greengrass client software service that runs on the edge device.

- The second part shows the other services that run on the AWS cloud, which are well-integrated with the AWS IoT Greengrass client software or runtime that is running on the edge device. This is the **data plane** of the AWS IoT Greengrass client software service that runs on the edge. In other words, the AWS IoT Greengrass client software that runs on the edge communicates or sends **IoT data or messages** with different AWS cloud services, such as AWS IoT Core and Amazon Kinesis, which run on the AWS cloud.

- The circle with the letter **C** shows the other external or third-party services that AWS IoT Greengrass client software can interact or integrate with using different AWS IoT Greengrass client software connectors.

There are two software distributions of core Amazon IoT Greengrass:

- **Amazon IoT Greengrass Core software**: This is the software binary that should be installed on the edge device. At the time of writing, there are two versions of Amazon IoT Greengrass Core software; the latest version is V2.

- **Amazon IoT Greengrass Core SDKs**: These SDKs are mainly to enable the Lambda function running in the edge to interact with Greengrass Core to exchange an MQTT message with the Greengrass MQTT message broker, exchange an MQTT message with other devices and other Lambda functions in the Greengrass group, interact with a local shadow service, access secret resources, and many more other features. The SDKs are provided in different programming languages such as Java, Node.js, Python, and C.

  Also, if you have the IoT edge devices running on Amazon FreeRTOS, then you have the Greengrass library (SDK) built-in with Amazon FreeRTOS. Otherwise (such as a different RTOS or embedded software running directly on the IoT devices), you can use these Amazon IoT Greengrass SDKs to interact with core Amazon Greengrass.

You can find the preceding software distributions and others in the **AWS IoT | Software** section of the AWS Console, as shown in the following screenshot:

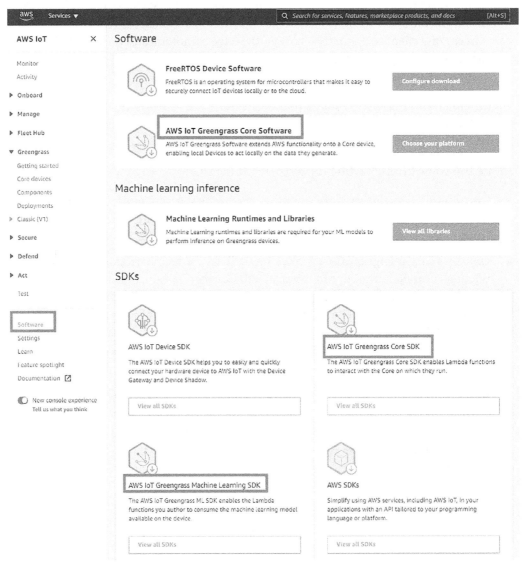

Figure 3.6 – AWS Greengrass software – from the AWS Console

As shown in the screenshot, there are other SDKs, such as AWS SDKs and AWS IoT Device SDKs. These AWS SDKs abstract lots of integration with different AWS cloud services.

The deployment architecture of an edge software runtime such as Greengrass usually follows a component-based or module-based deployment architecture. In other words, the provider of such an edge software platform gives developers the ability and flexibility to extend functionality running at the edge by deploying – remotely – new customized components in a plug-and-play deployment fashion to the edge software runtime.

As explained earlier, the control part of the edge software usually sits on the cloud side, while there is an agent usually installed and running on the edge software runtime to communicate with the cloud control management part for deploying and managing components on the edge runtime software.

Amazon Greengrass and Microsoft Azure IoT Edge follow this architecture. In Microsoft Azure IoT Edge, it is called a module, while in Amazon Greengrass, it is called a component.

An Amazon Greengrass component consists of the following:

- **Recipe**: This is a JSON or YAML file that describes the component by defining the component details, configuration, and parameters.

- **Artifacts**: These can be in the form of source code, binaries, or scripts that define the software of the component that will run on the edge runtime software. You can create artifacts from scratch, or you can create a component using a Lambda function, a Docker container, or a custom runtime.

- **Dependencies**: This defines the relationship between components that are run on the edge. You need this feature, for example, when you must install one component before another component, as the latter component depends on the former component.

In Amazon Greengrass, there are two types of components:

- **The nucleus**: This is the core software or minimum software that is run on the edge software runtime. It is responsible for managing the deployment, orchestration, and life cycle management of other components run on the edge software runtime. It also facilitates communication between AWS IoT Greengrass components locally on an individual device.

- **Optional components**: These are the configurable components that you can enable to run on an AWS Greengrass runtime, such as data streaming and ML inference.

Other concepts used in AWS Greengrass that you should be aware of are as follows:

- **Greengrass group**: This represents a logical collection of Greengrass Core software, IoT endpoint devices, and subscriptions belonging to that group. A Greengrass group must contain exactly one core. Currently, a Greengrass group can contain up to 2,500 devices. A device can be a member of up to 10 groups.

- **Subscription**: This represents the list of routing rules for MQTT messages. Each rule defines a source and a target, which can be an IoT endpoint device, a Lambda function, connector, shadow service, or AWS IoT Core. A source and a target communicate based on the topic or subject that was used to filter the messages. We will talk more about MQTT and topic structure later in the book.

Now, let's go briefly through some of the software components that come as built-in components with AWS Greengrass Core, examining their functionality and what they provide:

- **Local message broker**: Similar to AWS IoT Core, which is an MQTT-compliant message broker, AWS Greengrass Core has a local message broker component (Moquette) to allow and facilitate communication between endpoint IoT devices and the IoT edge, and also to enable the offline command and control operations between Greengrass Core and endpoint IoT devices that use the AWS IoT device SDK.

- **AWS Lambda at the edge**: This is a very good feature and was a game-changer in edge software, where you can have an event-driven AWS Lambda paradigm run at the edge as it does in the AWS cloud, so you can write your logic in a Lambda function (Python, Node.js, or Java) in the AWS cloud and then push it to Greengrass Core – as a component, as explained earlier – at the edge. The Lambda function will be triggered or invoked, with the messaging coming to the local message broker mentioned previously. To run Lambda at the edge, different subcomponents have been provided by Amazon Greengrass, such as Lambda launcher, Lambda runtime, and Lambda manager.

- **Local device shadow**: We will explain the device shadow service later when we talk about AWS IoT Core in the AWS cloud. Similar to what happens with the message broker and Lambda, the device shadow service has been pushed locally into the edge, and its main responsibility is to enable an IoT endpoint device to operate normally during intermittent connectivity and then synchronize the state with the AWS IoT device shadow service that is run on the AWS cloud when connectivity is restored.

- **Security**: The same authentication and authorization security model used in the AWS IoT cloud is pushed to the edge as well, so Greengrass Core supports TLS mutual authentication in both directions; one direction is between IoT endpoint devices and Greengrass Core and the other direction is between Greengrass Core and the AWS IoT cloud. Also, the authorization policy typically configured in the AWS cloud is pushed to Greengrass Core at the edge.

- **Local secrets manager**: Just like the AWS Secrets Manager service in the cloud, which is a feature added to the Greengrass Core to store, access, rotate, and manage locally secrets and credentials that are used in the local Lambda functions, the control plane of that component is the AWS Secrets Manager service in the cloud. So, you manage the secrets on edge devices through the AWS Secrets Manager service in the cloud.

- **Local resource access**: An AWS Lambda function run on AWS Greengrass Core can access local device hardware resources. This is great, as a Greengrass Core device can also have sensors and actuators attached to it, and using the Lambda function, you can read the sensor and actuator data and take further actions accordingly.

- **ML inference**: This is also a very good feature and was a game-changer as well in that domain – for example, running an ML model at the edge, bringing lots of business benefits in terms of a quick response to detected events, working in offline mode, and many other benefits.

  The ML model is typically trained in the cloud using Amazon SageMaker, which is a managed ML service in the AWS cloud, or it can be trained using an EC2 machine running other ML platforms and solutions such as TensorFlow, PyTorch, and Apache MXNet.

  The trained model is deployed to the edge, and the inference of that model, is done at the edge; data can still be sent back to the cloud to enhance the ML model's accuracy.

- **Connectors**: Greengrass connectors connect quickly and smoothly the edge devices to third-party services, AWS cloud services, or on-premises services.

  Connectors are pre-built functions that enable integration with services such as AWS Kinesis Data Firehose, Amazon CloudWatch, Amazon **Simple Notification Service** (**SNS**), ServiceNow, Twilio, and so many other services from AWS and other third-party SaaS providers.

- **Over the air updates agent**: This feature controls the deployment and configuration of Greengrass Core as it enables remote updates of Greengrass software, security updates, bug fixes, and component deployment. You can do a bulk update of many cores at the same time; also, you can check the status of updates from the AWS Console. During the updates, if something goes wrong, changes are automatically reverted.

- **Stream manager**: This is a kind of a local stream processing engine run in Greengrass Core. It processes the data streams coming from the IoT endpoint device locally and then exports them to an AWS cloud service such as Amazon S3, Amazon Kinesis Data Streams, AWS IoT Analytics, and AWS IoT SiteWise for further processing and analysis. This component is a fundamental component in doing real-time analytics and ML inference at the edge. It works by a data stream event being sent to an IoT edge stream manager component that is run on a Greengrass Core software runtime, following which a Lambda function hosting the ML model is triggered to evaluate the incoming data stream event against the ML model and return the output immediately. For example, assume you have a trained ML model in the AWS cloud for detecting the occupancy of a parking lot. The training is usually done by feeding the model with lots of data – in our case, the data is a set of images showing the occupied and unoccupied parking lot. Once the model is ready – that is, the model can detect whether the parking lot is occupied or not – then you deploy it. In our case, we deploy the model to the edge or to the AWS Greengrass Core runtime environment. In the end, the model is captured in the form of a Lambda function that runs locally in Greengrass Core. Now, in the runtime, a surveillance camera (the IoT endpoint device) will send a stream of images of the parking area to a smart parking edge device that runs AWS Greengrass Core. A stream of the images will land in the Greengrass stream manager component, which has a Lambda function associated with it – that is, Lambda will be triggered once the event comes in. The Lambda function holds the ML model and expects as input the stream event data (that is, the image), and the output will be either occupied or empty parking lots found.

- **Discovery service**: Devices at the edge need to discover the Greengrass Core group they belong to. You might have many Greengrass cores deployed on the edge; for example, on each floor of the building, you might have an edge device that has Greengrass Core installed on it. So, this feature basically enables the IoT endpoint device to discover the right Greengrass core to connect to.

- **Docker application manager**: This component enables AWS IoT Greengrass to download Docker images from public and private Docker image registries. If you create a custom component as a Docker image, this is the component that downloads the Docker image from a specified Docker registry and runs it.

Those are some of the built-in components of the Greengrass Core software. We do expect the list to increase in the future with the addition of more built-in components from the AWS side and more customized components built and provided directly by external third parties or through channels such as AWS Marketplace.

There are so many tutorials on the internet and specifically on AWS online documentation that provide step-by-step guides on how to implement the previously mentioned concepts and architecture of AWS Greengrass.

Now, let's try to summarize the important points that you need to consider as an IoT solution architect or designer during the design and implementation phases of the IoT edge:

- First, we explained the generic software components required in any IoT edge solution. You can use those generic software components in evaluating different available IoT edge solutions for your IoT production-grade solution and/or use those generic components as a starting point for implementing your IoT edge platform yourself, in-house, or with a vendor.

- We then explained one implementation of those generic software components and the services that are required in the IoT edge, which was the AWS Greengrass solution. It is the service provided by AWS covering the IoT edge solution.

- The AWS IoT Greengrass service has two parts – one part hosted on the AWS cloud and the other part being the software or runtime, which deploys and runs on the edge device at the edge side.

- Edge devices and IoT endpoint devices should be registered and managed by the IoT cloud platform. In the case of AWS, IoT endpoint devices and edge devices are registered in the AWS IoT Core service as IoT devices. We will explain more about AWS IoT Core later in the book.

- If the IoT endpoint devices on the field/ground are configured to communicate with the edge devices and not the IoT cloud, then the IoT edge devices will be in the middle. This is because from one end, they will communicate with endpoint IoT devices and from the other end, they will communicate and sync with the IoT cloud.

- SDKs, as they are known, simplify the development process and integration with different services in the cloud and at the edge. In the case of AWS, it provides lots of SDKs to enable integration with different AWS cloud services, AWS Greengrass, and AWS IoT Greengrass ML.

- The installation of AWS Greengrass is straightforward. You just need to follow the installation instructions provided by AWS for different hardware and operating systems – for example, there are instructions for installing on Raspberry Pi and instructions for installing on a Linux-based machine.

- In the AWS cloud, you create the different objects required for Greengrass and other components, such as registering edge devices and other endpoint IoT devices in AWS IoT Core. Also, from the AWS cloud side, you manage and control the Greengrass Core software and its components, which are run on an edge device.

The IoT edge or any other application workload that runs on the edge is a very important business and technical feature that every organization wants to implement.

Future trends all count on edge computing capabilities. Lots of effort and investment have already been made and will continue in the future to enhance the edge computing paradigm. I expect that in the future, edge technology will extend even further to the endpoint device itself – that is, your smartphone or IoT endpoint devices. At the time of writing, edge services are still too isolated from endpoint devices to avoid a device consuming too much power, resources that are limited and constrained, and so on, but who knows – in the future, this might completely change.

So, understanding edge computing, its different aspects, such as location, hardware, and software, and how it fits in with and solves IoT requirements is a very important and fundamental skill for any IoT solution architect and designer.

Summary

In this chapter, you learned about IoT devices and edge computing, covering different topics such as microcontrollers and how you can select the best-fit microcontroller or IoT device for your IoT project. Then, we covered embedded software, the need for an RTOS, the benefits of an RTOS, how you can select an RTOS platform for your IoT devices, and we explored FreeRTOS and Amazon FreeRTOS as examples of RTOS platforms. Finally, we covered the IoT edge from three perspectives – IoT edge location, IoT edge hardware, and IoT edge software. Regarding IoT edge software, we covered a generic IoT edge software platform, explaining the different components required for it. Then, we explained in some detail the Amazon Greengrass IoT software edge solution as an implementation example of IoT edge software.

Now, you will be able to identify and evaluate different solutions or options for IoT devices and the IoT edge.

With this chapter, we have completed all IoT aspects at the ground level where IoT devices and the edge will be deployed. The first three chapters mainly covered the sensing of data from the physical world and sending it to the backend IoT cloud for further processing and analysis.

In the next chapter, we will move toward another field and will continue with it till the end of the book – the IoT cloud. Going forward, we will cover topics related to the standard IT workload that runs on a cloud (either private or public) infrastructure and serves an IoT solution's different business requirements.

Section 2: The IoT Backend (aka the IoT Cloud)

The objective of Section 2 is to help you deeply understand the IoT backend layer. That layer is very important and is considered as the backbone of any large-scale and production-grade IoT solution.

You will learn about the different software components and solutions that the IoT backend layer has, such as IoT device management, IoT data analytics, and IoT platforms, to mention a few of the things explained in this section. We will also cover the infrastructure part of the IoT backend layer, the public and private clouds, VMs and containers, and Kubernetes.

This part of the book comprises the following chapters:

- *Chapter 4, Diving Deep into the IoT Backend (the IoT Cloud)*
- *Chapter 5, Exploring IoT Platforms*
- *Chapter 6, Understanding IoT Device Management*
- *Chapter 7, In the End, It Is All about Data, Isn't It?*

4
Diving Deep into the IoT Backend (the IoT Cloud)

In previous chapters, we mentioned the terms **Internet of Things (IoT) cloud** and **IoT backend** quite a lot. In those chapters, we were focusing mainly on IoT endpoint devices, IoT Edge devices, and the connectivity between those devices and the internet.

The goal of the previous chapters was to show the ability to sense the physical world and act upon the useful business operation data collected from the physical world through IoT devices, then through the different IoT connectivity options; such generated and collected IoT data is sent to the IoT backend operation and control systems, or IoT cloud in short, for further processing and analysis.

The *sensing* part is directed from the IoT devices toward the IoT backend systems, which follows the IoT telemetry design pattern. The opposite direction is from the IoT backend systems toward the IoT devices to act upon and control those IoT devices, which follows the IoT command design pattern.

Based on what has been explained so far, the IoT backend or IoT cloud might seem straightforward but in reality, and in large-scale, production-grade IoT solutions, it is not that simple, as most IoT solution project efforts, investments, innovations, and business values are heavily dependent on and derived from the quality and strength of your IoT backend cloud. So, the more scalable, reliable, secure, performant, and resilient your IoT cloud or IoT backend is, the more robust and profitable your IoT solutions will be, as well as delivering the desired business outcome.

The IoT cloud topic can be broadly categorized into two subtopics. One subtopic covers **cloud infrastructure** and the different services and capabilities provided for building the IoT cloud infrastructure, which cloud you should use to host your IoT backend – the public cloud, private cloud, multicloud, or hybrid cloud model – and what the benefits, trade-offs, and challenges of each cloud option are. We will discuss not only which cloud option but also which cloud offering you should use, that is, **Infrastructure as a Service (IaaS)**, **Platform as a Service (PaaS)**, or **Software as a Service (SaaS)**. The other subtopic covers the different **IoT applications and solutions** that will run on top of the IoT cloud infrastructure.

In this chapter, we will focus mainly on the infrastructure part of the IoT cloud and the different services and capabilities required in building the IoT cloud. The IoT applications and solutions will be covered in subsequent chapters, for example, the **IoT device management solution**, **IoT device gateway**, **IoT message broker**, **IoT rule engine**, **IoT application dashboard**, and **visualization**.

In this chapter, we will cover the following topics:

- The IoT cloud
- Kubernetes (the container orchestrator)
- AWS cloud overview

The IoT cloud

Before we go into the details of the **IoT cloud**, let's first understand the difference between a data center and the cloud.

Data center and the cloud

Traditionally, organizations hosted their IT workloads and systems either in data centers that were on-premises (in-house) or off-premises (off-house) and supported by internal infrastructure teams or third parties or in external data centers supported and maintained by external third parties altogether.

Running and maintaining a data center is not an easy task; if you run a data center on your own, then you will need to deal with things such as finding suitable buildings or data center facilities to host your servers and network infrastructure, power supply and consumption, physical security, and governmental regulations.

From a business point of view, running and maintaining a data center is a business domain in itself. So, the question will be, as a company working in the agriculture, transportation, health, or any other business domain, do you need to run another business domain, such as data center management, besides your original business domain or will it be better if you just focus on your business domain and lease or hire another company that is an expert in running the data center domain?

From our point of view, the answer to the preceding question is: companies should focus on and innovate their main business domain and leave the data center management domain to specialized data center providers. However, we shouldn't neglect the history and the huge investment that has already been spent on organization-owned data centers. This is where organizations with data centers that they own enter the debate of whether to migrate their workloads to public cloud providers, keep investing in their private data centers, or use a hybrid approach, which is some of the workload deployed into the public cloud(s) and other workloads to be continued on their private cloud(s). On the other side, most start-ups and innovative companies are born on the public cloud; in other words, they use the public cloud from day 1, leaving them to focus on their business ideas and how to differentiate themselves in the competitive market we live in.

Before we continue further, don't get confused between the terms cloud and data center. Both are referring to almost the same thing, which is the physical building where the racks, servers, switches, routers, storage, power supply, and other elements are hosted and maintained. Cloud or cloud computing paradigms offer or enable much easier and more convenient provisioning and on-demand access to different cloud resources, such as compute, storage, networking, applications, and services, in public and private clouds as well.

With a click of a button in a **console**, in an **API call**, or through the **Command-Line Interface (CLI)**, you can provision a complete virtual data center with all the resources you require without the need to physically provision servers, switches, routers, and so on, or have any human interaction at all.

In the next section, let's see the characteristics of the cloud, in other words, what differentiates cloud computing from a traditional data center.

Cloud characteristics

There are some characteristics that identify the cloud, including the following:

- **Resource self-service management**: As a consumer of the cloud, you can provision and manage different cloud resources on your own using your own convenient way or an engagement channel. Usually, cloud providers offer different engagement channels or clients to interact with their cloud resources remotely, such as the following:

    - **Portal/Graphical User Interface (GUI)/console**: This is the human interaction interface and is usually used for easy and small tasks or quick troubleshooting, but not for automation tasks, such as creating cloud resources through the GUI, which will take lots of time and effort. Ideally, such types of tasks are not done through the GUI.

    - **Application Programming Interface (API)**: This is the system interaction interface and is usually used for automation and system integration. Behind the scenes, cloud providers have a very rich set of APIs that covers all operations of all cloud resources that they offer. The cloud-provided console or portal (and cloud-provided CLIs and SDKs; see the following points) uses or is built on top of those cloud provider APIs. You can build your own cloud portal or integrate some existing portals or systems into your organization with the cloud using the cloud-provided APIs.

    - **CLI**: This is another system interaction interface and is used mainly for automation scripts or command-line scripts that use cloud-provided APIs behind the scenes.

    - **Software Development Kits (SDKs)**: Cloud providers usually provide different SDKs in a different programming language (for example, Java, Python, or .NET) to simplify the development and integration with their cloud services. SDKs use cloud-provided APIs behind the scenes.

The list can continue with other channels, such as a mobile app, but the idea is very simple: any interactive channel will use cloud-provided APIs to integrate with the cloud. Some cloud providers provide fancy channels, such as a mobile app, already and others don't. If you want to build additional clients, then simply use the cloud provider APIs to do it.

- **Resource sharing**: A cloud computing model is by default a multitenant model. In other words, cloud providers usually have a massive set of cloud resources available, so they share or divide those resources between different cloud consumers. This was the idea that triggered the public cloud computing industry. Amazon.com is one of the biggest e-commerce platforms on earth. Amazon provisions lots of servers during seasonal and occasional events, such as Black Friday and Boxing Day, as the traffic increases during those times. After those events have finished, the additional cloud resources that are added are not utilized and become idle until the next year. As this is a waste of money and resources, Amazon at that time came up with the great idea of leasing or renting these additional cloud resources to other companies to create a win-win situation, where Amazon gets back the money they invested into those additional servers and hardware(as well as making a profit). The companies buying those resources will get the benefit of a leasing model where they don't need to worry about buying such resources and running into the same issue we just explained about Amazon's idle resources. Now, companies pay for what they use only, and they can shut down those resources anytime without any contractual issues.

- **Resource elasticity**: This characteristic is what specifically differentiates the cloud from traditional data centers. The idea of scaling up or down (or scaling in or out) either automatically or manually is a great feature of the cloud. In this model, you don't need to worry about the initial investment and the capacity plans for your infrastructure.

  With cloud computing, you can start small and scale up (or scale out) when needed and you can scale down (or scale in) and save resources when needed.

  Usually, cloud providers say they have an unlimited set of resources for their cloud consumers, but the reality is there is for sure a limit or a cap on the number of cloud resources they have. This is why there is what is known as a soft limit and a hard limit. With a soft limit, you can increase – to some level – your allocated quota but you need to discuss this with the cloud provider first. With a hard limit, you can't increase or go beyond your quota.

- **Resource usage metering**: In the cloud, the default billing model is a pay-as-you-go model, or pay for what you have consumed, thanks to the resource usage metering mechanism behind the scenes that allows cloud providers to charge cloud consumers on a per-hour, per-minute, or even per-second basis. This is also another characteristic that differentiates the cloud computing model from the traditional data center model since in traditional data centers, you don't need to worry about usage as you already own and have paid upfront for the infrastructure; so, whether you fully use it or not, it doesn't matter.

Cloud computing is sort of a managed service and managed services are all about abstractions. As a managed service provider, the more you abstract from the service consumers, the more you offer your managed services to your customers.

In the next section, let's go through the list of **cloud service offerings** and see the level of abstraction at each level.

Cloud service offerings

The services offered by cloud service providers can be split into three well-known categories:

- **IaaS**: In this cloud service offering, the cloud provider abstracts away from the cloud consumers things such as data center facilities, networking, data storage, host servers, and hypervisors. The cloud consumer takes responsibility for the virtual network, **Virtual Machines (VMs)**, operating systems, applications, middleware, and data.

    This model gives much more control to cloud consumers in comparison with other cloud service offerings mentioned later.

    For example, you launch a VM where you can do or install whatever you want on it and you have full control of that VM.

- **PaaS**: In this cloud service offering, the cloud provider abstracts away more features and tasks from the cloud consumers – things that are already abstracted in the IaaS model, plus taking care of the virtual network, VMs, operating systems, and middleware or runtime. What is left to cloud consumers is mainly the application and data.

    In this model, cloud consumers have less control; they control only their application and data and nothing more.

    For example, AWS Beanstalk is a PaaS offering as AWS takes care of everything and gives cloud consumers access to an up -and -running environment of selected runtime or middleware platforms. For example, if you want to deploy a Java application, you simply choose Java as the runtime of AWS Beanstalk and AWS will provision the required VMs, operating systems, and Java runtime to deploy your Java application behind the scenes, as well as handling the scaling, monitoring, and security of that environment.

- **SaaS**: This is the most abstracted cloud service offering provided by cloud providers where the cloud providers take care of everything, and cloud consumers are just users or consumers of those SaaS services.

For example, Gmail, provided by Google, is considered a SaaS service where Google takes care of the whole thing and you access and use that service over the internet.

There have been some other cloud service offerings launched recently following the **as a Service (aaS)** model, for example, **Container as a Service (CaaS)**. CaaS is a new cloud offering that is dedicated to running and hosting container-based applications. There are some debates about CaaS and whether it is a new category in itself or whether it is just an additional runtime (that is, a container runtime) option that should be offered under the existing PaaS model (other PaaS runtimes, such as Java, Python, and .NET). There are some cloud providers that offer container deployment or runtime options in their PaaS offerings. While other cloud providers make a dedicated model (CaaS) for containers, as with the container runtime, you can really run anything (Java, Node.js, Python, and so on) so that needs special treatment.

There are other models, such as **Monitoring as a Service (MaaS)** and communication as a service. Note that communication as a service differs from container as a service, which we explained before, and in the rest of the book, when we refer to CaaS, we mean container as a service, until we come to **XaaS (Anything as a Service)**.

Now that we know some details about the different cloud offerings, the question will be, which cloud offering should I choose for my IoT backend cloud?

Well, the answer will depend on many factors, including the following:

- You need to know what your (or your company's) strategy is for the IoT cloud; is it okay to go with the public cloud or are there some concerns at the wider company level with using the public cloud? What about hybrid cloud or multicloud? (More about those deployment models and which one to choose will be covered in the coming section.)

- You need to know what your (or your company's) strategy about the IoT backend from a portability point of view is. In other words, are you (or your company) okay with being locked in with a cloud provider or do you (or your company) have an avoid vendor lock-in strategy in place? In other words, are you looking to move your workload simply and smoothly between different clouds or not?

    If portability is a must, then IaaS is the best choice for you as you have full control. You will have VM services with every cloud provider, and they provide tools to migrate VMs from on-premises or another cloud provider to their cloud and vice versa.

If portability is a nice-to-have requirement (that is, an optional requirement), then you can consider other options, such as PaaS and SaaS, especially when you have other critical requirements, such as launching your (or your company's) IoT products or solutions quickly to market. For such a requirement, your best option will be a PaaS or SaaS model to quickly launch your products without worrying about installation, support, scaling, and so on.

- What is the architecture of the workload you need to deploy into your IoT cloud? Is it based on a legacy architecture where you need a specific hardware or operating system to run on top of? Or is the workload modern, cloud-native, and infrastructure-agnostic?

  If the IoT workload is based on legacy architecture, then IaaS will be the best option with some effort of migration and enhancement for those legacy workloads to be able to run on a cloud environment.

If the IoT workload is cloud-native or cloud-ready (that is, it can be moved to the cloud), then the other options could be considered in the selection process, such as PaaS, CaaS, or even SaaS.

- Last but not least, is the IoT project a greenfield project where you will have to start from scratch so that you are 100% sure that all of your workloads will follow a cloud-native approach, or is the IoT project a brownfield project where you will have a mix between new and old workloads and the migration of the old workload to a new, modern workload is not clear or defined?

Ben Horowitz said, "There is no silver bullet. There are always options and the options have consequences."

To conclude this section, we cannot say which option is best for your IoT cloud as it all depends on the different, dynamic factors you might have in your IoT project. However, our view is that **CaaS** is the best option if you have a containerized IoT workload. Why we say that is because of the following:

- Most container technologies (for example, Docker and Kubernetes) are open source so you will avoid challenges such as vendor lock-in, licenses, and portability.

- You can move a containerized workload anywhere. You simply need Docker Engine (if you use Docker containers) and Kubernetes (if you use it for container management and orchestration). With that, you will overcome the challenge of portability.

- You can use container-managed services provided by cloud providers and you are still free from the vendor lock-in challenge. Why? As the core thing in containerized applications is the container images, and this is what you own and build, you can now use a cloud provider to run those containers using their container-managed service, and tomorrow you can move to another cloud provider to run those containers using their container-managed service as well. So, in short, as long as you own the container images, you are free to select any cloud provider-managed container services, or you can even follow a **Do It Yourself** (**DIY**) approach with any public or private cloud provider using their IaaS.

We will explain container technology in a separate section as it is a very important topic for IoT cloud infrastructure.

Now, in the next section, let's cover the different cloud deployment models.

Cloud deployment models

In the previous sections, we explained the cloud and what characteristics define or differentiate it from traditional data centers. In this section, we will cover the different cloud deployment models.

Private cloud

As the name suggests, this type of cloud is exclusively owned and used by a single organization. Within that organization, you can have many consumers of that cloud, such as sub-organizations or organizational business units.

In a private cloud, if the organization that owns the private cloud is buying or renting data center facilities to host their infrastructure racks, servers, switches, routers, storage, and so on, the data center facility or building is typically supplied with power, connectivity, physical security, and so on.

To run a cloud, you need cloud computing software to manage the computing, networking, storage, and other cloud services. The most common and widely used open source and free cloud software is **OpenStack**.

Here are some benefits of a private cloud:

- **Flexibility and control**: In a private cloud, you own everything about the cloud so you have a lot of control and flexibility. For example, you don't need to wait for other cloud provider roadmaps to have some features that you need in your cloud; you can simply go and do it.

- **Security and privacy**: This will usually be the first answer you hear when you ask someone why they are choosing the private cloud over the public cloud. The belief of having the customer data secured and not traversing by any means outside the boundary of the organization makes the private cloud seem the best option for such data privacy and security concerns.

- **Cost**: This is a debatable one, as some people see private clouds as cheaper than public clouds, but it really depends on the size of the organization and the nature of the workload that is running in the cloud. For example, is the workload running in the private cloud running consistently for a long period or infrequently? How frequently do you need to refresh the hardware that is running in that private cloud? In some cases, a public cloud is cheaper than a private cloud.

The cost model is different between private clouds and public clouds. Private clouds use a **Capital Expenditure** (**CAPEX**) model, which is the investment the organization puts in. In other words, in the private cloud model, organizations buy the assets and own them, while in the public cloud model, organizations use an **Operating Expense** (**OPEX**) model, which is a day-to-day expenses or monthly expenses kind of model. In other words, organizations in the public cloud model do not own the assets or the resources; they just rent them and pay for what they use.

Here are some of the private cloud challenges:

- **Scalability and footprint**: The size of the private cloud can't be compared with hyper-scale public cloud providers, such as Amazon, Microsoft, and Google. Public cloud providers have massive data centers across the globe, so scalability in the private cloud is a big challenge.

- **Maintenance and support**: A private cloud requires lots of maintenance and support. Usually, there are dedicated teams or departments to look after the maintenance of private clouds, which adds additional costs to the private cloud and makes it more challenging.

- **Cost**: As we discussed in the benefits list, this is a debatable point. Some organizations see private clouds as very expensive compared to the public cloud model as the cost will not only cover the data center facilities, hardware, and so on, but it also requires staff to manage and look after that private cloud.

> **Note**
> Private doesn't mean the organization has to manage it by itself. An organization can subcontract the job of managing the organization's private clouds to a third-party vendor on their behalf. Also, it doesn't mean the cloud is hosted on-premises; it can be deployed off-premises with third-party vendors.

Public cloud

As the name suggests, this type of cloud is shared between different public consumers from different business domains. The public cloud providers provision and manage cloud resources for different consumers in a multitenant fashion.

Public cloud providers usually follow the shared responsibility model when it comes to security in the public cloud. In the case of an IaaS offering, for example, it will take care of the data center's physical security (for example, access to data centers) and cloud infrastructure security (for example, compute, storage, and networking), while the customers or consumers of the public cloud will take care of the remaining aspects, such as customer data privacy and security, operating systems of VMs, applications, platforms, and identity and access management.

If it is a PaaS offering, then public cloud providers will take more responsibility in that security shared model, whereas in a SaaS offering, public cloud providers take care of everything.

Public cloud providers usually provide a massive set of security controls and tools to help customers in that shared responsibility model.

There are many public cloud providers available on the market, such as **Amazon Web Services (AWS)**, **Microsoft**, **Google**, **Oracle**, **IBM**, and **Alibaba**. In this book, we broadly use AWS for our examples and references for different IoT services and solutions.

Here are some of the benefits of the public cloud:

- **Scale and footprint**: In the public cloud, you can scale easily. Usually, you start with a small scale and you can scale easily later if your business grows or if you need to launch a business in another country or region. This is the big differentiator between the public cloud and the private cloud. In the private cloud, if you want to scale more, then you need to make an order for purchasing additional hardware, servers, and so on, which is usually time-consuming, costly, and a task that requires a lot of effort.

This scale of the public cloud also helps in setting up a proper deployment architecture for your workload. Features such as redundancy, resilience, performance, and high availability have become easy to provide with such a massive set of cloud resources that public cloud providers have and can offer. You'll need to pay more for the additional resources for your deployment, but this is usually an architecture trade-off discussion with the project stakeholders; in other words, if the business needs the IoT backend to be more resilient, highly available, and geo-redundant, then the cost to achieve that will be high. In other business contexts or projects, where the business might decide to use a minimum budget or low cost for the IoT backend, then you might go with options such as single-region redundancy (not geo-redundant) or you might use fewer resources; hence, availability and resiliency levels will reduce or get impacted because of that limit in resources.

If you think about hyper-scale public cloud providers, such as Amazon, Microsoft, and Google, they already had massive data centers across the globe before the cloud computing era. What they are doing now is improving the efficiency of such existing data centers and adding more. This is why scaling with any of those providers is a very easy task due to the massive pool of resources they have.

- **Innovation**: The rule is, if you lower the cost of experimentation, then you can innovate more. The public cloud costing model is pay as you go or pay for what you use. So, you can simply do proofs of concepts in a week or two, for example, and then you shut down the resources that you have used and stop the billing for those resources accordingly. The pace of innovation in the public cloud is very fast. Public cloud providers also participate in that innovation race by providing the proper infrastructure, tools, and cutting-edge technologies to cloud consumers to help them innovate faster in their business domains.

- **Reduced Total Cost of Ownership** (**TCO**): With the public cloud, you don't need to worry about the cost and maintenance of data center facilities, hardware, licenses, support, and so on. Public cloud providers take care of all of this and you just focus on what matters for your business.

- **Agility and time to market**: With public cloud services, especially *PaaS* and *SaaS* services, you launch your product very fast as most of the heavy-lifting work is already taken care of by the public cloud provider. For example, in the managed database services provided by a public cloud provider, the public cloud provider will take care of the installation, maintenance, support, backup and restore, resiliency, and performance of the database engine. These tasks, if you do them yourself, will take lots of time and effort and will be done with less efficiency compared with public cloud-managed services.

- **Pay-as-you-go/use model**: This is a very attractive feature of the public cloud deployment model, especially for **Small and Medium Enterprises (SMEs)** and start-ups. With a pay-as-you-go model, you hit the ground running from day 0 and don't have to worry about capacity planning and investment in purchasing hardware, licenses, long-term contracts, and so on.

- **Workforce productivity**: With public cloud-managed services, you focus on what matters for your business and leave non-business tasks to the public cloud provider. For example, if you want to launch a Hadoop cluster for some big data and analytics solutions, do you care more about installing, maintaining, and supporting the Hadoop cluster infrastructure or about the business analytics jobs and use cases you are looking for? If you care more about the business analytics jobs, then you can go with public cloud-managed Hadoop services and launch the Hadoop cluster in a few minutes with a few clicks in a console or through an API call. So, in a few minutes, you are ready to go with your analytics jobs. Awesome, isn't it?

Here are some challenges that come with public clouds:

- **Loss of control and customization**: If for whatever reason you need to have control of the infrastructure and hardware level, then the public cloud is not the best option for you or your organization.

- **Security and privacy**: Despite the fact that public cloud providers provide a massive set of security controls and tools for you to secure your customer data and applications in the cloud, there is still a feeling of worry about putting your customer data outside your boundaries or control.

- **Compliance issues**: Despite the fact that public cloud providers cover or comply with lots of compliance requirements, such as **PCI-DSS**, **HIPAA/HITECH**, **FedRAMP**, **GDPR**, **FIPS 140-2**, and **NIST 800-171**, there is still a wide array of legal and industry rules that organizations have been asked to comply with to ensure security and proper maintenance that make the private cloud the best option in that case.

Hybrid cloud

The hybrid model is a mix between the public and private models. In other words, you have some workloads that run in the private cloud and some other workloads that run in the public cloud.

Public cloud providers understand the importance of this model for large organizations and enterprises, hence they have started to offer their public cloud services to run on the customer's private cloud. For example, if you run an AWS EC2 machine (VM) in the AWS cloud, you can run the same in your private cloud with the same experience you get in the AWS public cloud.

In other words, public cloud providers host and manage their racks of servers on customers' private premises to give customers the same experience as the public cloud and avoid the challenge of data privacy and security at the same time.

The AWS hybrid cloud solution is called **Outposts**, Microsoft's solution is called **Azure Stack**, and Google's solution is called **Anthos**. Those hybrid cloud solutions provided by the hyperscale public cloud providers are clearly a down-scale version of their public cloud versions. In other words, not all services that run in the public cloud will be supported in those hybrid cloud solutions, but AWS, Microsoft, and Google keep improving and adding services to those solutions over time.

Here are some benefits of the hybrid cloud model:

- When it comes to decision making, hybrid is always a good option, as it combines the best of both worlds, the private cloud and the public cloud.
- Strategically, hybrid is the less risky option as you are not putting all your eggs in one basket.
- There is no waste of investment that has been made already in the private cloud.
- It is a more flexible option for covering things such as regulations, legal requirements, and infrastructure customization if needed.

Here are some challenges that come with the hybrid cloud:

- **Costly**: As you operate different clouds with different costing models, staff, and so on, it gets costly.
- **Operation overhead**: You need to maintain more and different cloud environments, which adds overhead to operation teams. Even if you use PaaS or SaaS offerings of the public cloud, you still need the operation team to look after that cloud workload as well.

There are other cloud models, such as the community cloud model, which is the same as the private cloud model with one exception, which is that the community cloud model serves a set of organizations (not just one organization) that share the same concerns or business domain. For example, a hospital community cloud is a private cloud for a set of hospitals in a country or region.

Now that we have learned about the different cloud deployment models, the question will be, *where should I deploy the IoT backend workload*: in a private cloud, public cloud, or hybrid cloud?

As we mentioned previously, there's no silver bullet; there are options and trade-offs for each. There are some points you need to consider before answering that question, such as the following:

- *What is your organization's cloud strategy?* This is a very important question you should ask and you should know its answer first as you might have a clear strategy to not use the public cloud for whatever reason, or you might find the opposite, that the organization's strategy is to use the public cloud and migrate from a private cloud to the public cloud. Or, you might find the hybrid cloud model is the recommended model in your organization.

  If your organization's cloud strategy is to use the private cloud only, then there's no further analysis needed, and you must use and deploy your IoT backend workload into a private cloud.

  If your organization's cloud strategy is okay with the public cloud, then we recommend deploying an IoT backend workload into the public cloud due to the benefits we mentioned earlier.

- *What is the type of the IoT backend workload architecture?* Is it a legacy-based architecture or does it follow a modern and cloud-native architecture? Is it a greenfield project or a brownfield project? If it is a brownfield project, what is the percentage of new IoT backend workloads that needs to be built and deployed compared to the existing workload? What is your organization's time-to-market strategy? Do you need to launch the product quickly or do you have time?

  The more you are leaning toward things such as an IoT backend greenfield project and modern and cloud-native architecture – the majority of IoT backend components are new to design and build, are agile, and have a quick product launch -to -market time – the more likely it is you should select the public cloud (assuming the organization's cloud strategy allows that). If you've answered the opposite, you should go with the private cloud or hybrid cloud (assuming your organization's cloud strategy allows that).

Now we understand the differences between the cloud and traditional data centers, the definition of the cloud and its characteristics, different cloud service offerings, and different cloud deployment models. In the next section, we will talk about a special platform or compute engine that is important for every IoT solution architect and designer to be aware of: Kubernetes. We will discuss its usage and the benefits it provides to your IoT backend cloud.

Kubernetes (the container orchestrator)

As stated earlier, from our point of view, Kubernetes is the best compute engine option to run your IoT backend cloud workloads, for many reasons. To mention a few, it is open source (created initially by Google, and then Google open sourced it), it has huge community support, it has massive ecosystems, solutions, and tools built around it, it is widely adopted, tested, and used by so many large organizations across the globe, and it is supported and offered as a managed service by almost all public cloud providers (AWS, Azure, Google, Oracle, and so on).

But what attracts us the most to Kubernetes is its portable nature, which is driven by it being open source . For example, you can have a Kubernetes cluster running on a private cloud on VMs or bare metal, on a public cloud IaaS, or on a public cloud CaaS, you can have many Kubernetes clusters running in different clouds (private or public), and you can control all of those clusters here and there using platforms such as **Rancher** or **OpenShift**.

Before we talk about Kubernetes, let's talk about **containers** first. But note that this book is not about containers or Kubernetes, hence we will cover both at a high level, mentioning only what is required for IoT solution architects and designers to be aware of.

Containers

When virtualization technologies (hypervisors) were introduced, they fundamentally changed the way the computing domain works. Until now, virtualization technologies were considered the core technology in private, public, and traditional data centers.

The idea of virtualization is brilliant, as you can virtualize, split, or share compute resources, such as the CPU and memory, of one physical host server/machine with multiple virtual servers/machines. A hypervisor is the kind of software that enables the concept of virtualization. There are many hypervisor distributions, such as **VMware ESX** and **ESXi** and **Microsoft Hyper-V**.

Before virtualization, if you had a three-tier application architecture, for example, and you needed to distribute the three tier layers (that is, presentation, application or business, and database), you would need a dedicated physical server to host the presentation layer, a dedicated physical server to host the application or business layer, and a dedicated physical server for the database. We also haven't mentioned redundancy, high -availability, and resiliency aspects, which will require at least two servers in each layer and a load balancer in front of those servers for high availability.

We are talking about one web application. You can imagine what the data center server farm would look like with such a massive number of physical servers. Think about the space required to host those physical servers and the power and cooling required to run the data center. Or even think about the physical server resource utilization and what percentage of the resources would be idle versus active during the server uptime. This is why virtualization was a great invention to overcome the preceding challenges of physical servers. Now, with virtualization, you can host the same three-tier web application that earlier required about 20 (for example) dedicated physical servers in 2 or 3 physical servers; that is, those 2 or 3 physical servers would host 20 virtual servers.

Virtualization virtualizes the host server resources to host multiple VMs. You can install and run whatever operating systems you want in those VMs. Those operating systems on those VMs are called guest operating systems. The operating system installed on the physical server or machine is called the host operating system.

The container technologies take that virtualization concept further to virtualize the operating system itself, so you can run different applications in the same operating system with full isolation between those applications.

Now, with the three-tier web application example we mentioned earlier, you can deploy that application into one VM – for example – and that VM will run or host 20 isolated containers and applications running on those containers as if they are running in 20 VMs or physical servers. That is awesome, isn't it?

The container concept was not new or modern, as the concept emerged from different Linux kernel features and technologies, such as **control groups (cgroups)**, **Secure Computing (Seccomp)**, **Security-Enhanced Linux (SELinux)**, and **Linux namespaces**. But Docker (the company) provides a rich set of tools and platforms that shape those Linux concepts and technologies into a proper, easy-to-use-and-package Docker container concept.

There are many container technologies and runtimes, such as **rkt** and **Podman**, but **Docker** is the most widely used container technology on the market, hence we will use it as a reference going forward.

Docker container technology has many components, such as **Docker images**, **Docker daemons**, **Docker Compose**, and **Docker Engine**.

Docker Engine is like the hypervisor used in the case of VM deployment. Once you install the hypervisor software, you can virtualize the host physical server and run the VMs on top of it. Docker Engine uses the same idea; once you install Docker Engine, you can run Docker containers on top of it. This is greatly beneficial in terms of the portability of your applications. You just need Docker Engine to be installed; you can install Docker Engine on a VM, on a physical host operating system, or even on bare metal.

If you write a microservice or monolith software service that has basic **Create, Read, Update, and Delete (CRUD)** operations and do some business logic processing, why would you deploy that service in a VM with a guest operating system that might take 5 minutes to start as it needs to load a bunch of different software libraries and drivers that might not be needed at all for running this microservice application? To clarify further, why would this microservice be concerned with loading printers, serial ports, or other peripheral software drivers and libraries?

From a development point of view, you write the software as usual and then package the software with all the required configurations, libraries, and other dependencies into a single Docker image. You can then deploy that image anywhere; you just need Docker Engine installed to run the Docker image. Once the Docker image is started, it is now called a Docker container.

So now, with Docker images, we have a new way to package or distribute the software. Previously, we used to distribute software in the form of TAR, WAR, JAR, EAR, DLL, .exe, and other file formats.

There are so many benefits of using containers when compared with VMs, such as the following:

- **Resource efficiency**: Containers use or share the operating system, so they use very little resources compared to VMs, which require a dedicated operating system.

- **Portability**: Containers truly follow the write once, run anywhere concept, so they provide a great portability option.

- **Scalability**: Containers fit perfectly with stateless apps, where scalability is up and down, and containers start and shut down very quickly compared to VMs.

- It is the perfect deployment model for **microservice-based** and **cloud-native applications**. Microservices are usually 100% stateless, have a small footprint, and are independent or isolated, and all those features fit and align very well with the container technology.

There are some challenges with container technologies, though the most obvious one is **security**. Despite container technologies providing the required separation and isolation between containers, they still run on a shared operating system. So, if one container is vulnerable, it could impact other containers running on the same operating system. This issue is not there in the case of VMs. If one VM is impacted, it will not impact other VMs, even if they're running in the same host server, as the virtualization is done at the hardware level (CPU, memory, and so on) and not at the operating system level.

The question that arises now is, which technology should we use: VMs or containers? Again, as mentioned previously, there's no silver bullet. There are options; in some cases, containers fit very well, such as microservice-based and cloud-native apps, while in legacy-based apps or apps that require access to operating system functionalities, VMs will be the best fit in that case.

There is a third option, which is to combine the best of both worlds, VM and container technologies, which is **MicroVM**. This is a lightweight VM that can start as quickly as containers and at the same time is fully secure and isolated (being a VM). You can also run container-based apps on top of a MicroVM.

In our view, if you build a new IoT backend from scratch or already have a well-designed IoT backend that could be containerized, then go with container technology; else go with MicroVM or VM technology.

Containers are great technology, but you need a container management platform or container orchestration engine to see the true greatness and value of containers. You need a platform to manage containers at scale. Do you know how many containers Google uses per week for running their Google search engine? Google it and you will see they have billions of containers running per week, so they clearly use a very strong container orchestration engine to manage that scale.

You also need the container orchestration engine in your IoT backend cloud. You will obviously not have the same scale as Google but in the end, you need to manage your application containers properly even if you run 100 containers or less.

On the market today, there are lots of container management and orchestration solutions, such as **Kubernetes**, **Apache Mesos**, and **Docker Swarm**.

In the next section, we will talk about Kubernetes only as it is the market leader and the most widely used and adapted container orchestration engine on the market as of the time of writing.

Kubernetes overview

Kubernetes, or K8s for short (as there are eight characters between the letter K and the letter s in the word Kubernetes), is an open source container orchestration engine.

Here are some of the features of Kubernetes:

- **Auto and manual scaling**: In Kubernetes, you can scale up and down the number of containers running manually or automatically based on CPU metrics.

- **Self-healing**: Kubernetes automatically restarts and replaces failed containers and reschedules containers into new worker nodes when their initial worker node dies or fails. Also, it manages and monitors health checks of containers and responds accordingly, which means that if the health check fails, it will kill the unhealthy container and replace it with other healthy containers, and so on.

- **Service discovery**: In an environment where you support autoscaling or horizontal scaling in general, containers will come and go. So, containers, or to be more accurate Pods (we will talk about that concept later), will have new IPs every time they start, so you need a centralized service registry (Pods or containers register the IPs that they got from the network layer when they start into a centralized service registry).

  Centralized service discovery, which is responsible for discovering the currently running Pods/containers, typically works and integrates with services such as a load balancer. For example, how does the load balancer know about the new containers/Pods attached to the load balancer targets and to send or stop sending traffic to those new or killed Pods/containers, respectively? This is done with the help of a service discovery service. That service discovery service comes built into Kubernetes.

- **Secret and configuration management**: Kubernetes externalizes the application configuration and the secret management for the containers running inside the cluster. So, the container running in Kubernetes reads the required configuration from the external configuration management system. Instead of storing application configuration or credentials inside the container in a property file or even hardcoded with the software running inside the container, you rather create

- A **ConfigMap Kubernetes object**, which holds this configuration outside the container process, which gives lots of flexibility in container deployment scenarios (for example, you can have ConfigMap for a development environment, which differs from ConfigMap for a production environment, and so on). The same applies to **Secrets**, things such as the API secret key and passwords. Ideally, you store such credentials in a **secure vault system**, not in plain text, and Kubernetes provides a **Secret object** that acts as a vault for storing such credentials.

There are many other features available in Kubernetes, but we can't cover them all in this book, so we will try to give only the high-level details that are beneficial or required for the job of IoT solution architect or designer.

In the next section, let's cover the Kubernetes architecture.

Kubernetes architecture

A Kubernetes cluster consists of two types of nodes playing different roles, the first being Kubernetes master nodes or the Kubernetes control plane. This type of node manages and controls the Kubernetes cluster resources and objects that are deployed into worker nodes. Worker nodes or data plane nodes are the second type of nodes in a Kubernetes cluster. Actual containers run on the worker nodes and are controlled by Kubernetes master nodes.

The master nodes could be one node or one machine, but for high availability, it is usually recommended to run master nodes in at least three nodes. Also, you can have one worker node or many worker nodes, but it is usually recommended to deploy your containers in at least two different worker nodes for the sake of high availability and resiliency.

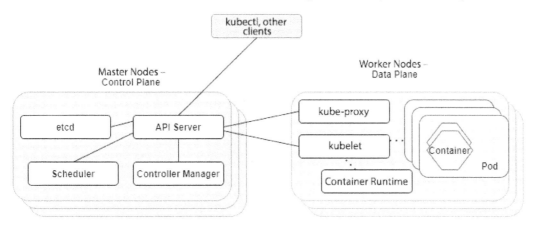

Figure 4.1 – Kubernetes cluster architecture

The master node has or must have the following components running:

- **API server**: You can say the Kubernetes API server is the brain of the Kubernetes cluster, as every request sent to the cluster to do something will have to first land in the Kubernetes API server and then the API server will decide what needs to be done.

Note that Kubernetes uses a declarative model, which means the client of the cluster just asks Kubernetes to create a Pod. It is then the Kubernetes master node's responsibility to create or schedule that Pod in the available worker node and do other tasks required for completing that request. Client requests or APIs (REST) come to the Kubernetes API server as it is the main entry point for the Kubernetes cluster.

The API server is responsible for the authentication and authorization of client requests.

- **etcd**: etcd is an open source distributed key-value store and is a critical component in Kubernetes architecture as it is used for service discovery, configuration management, and holding a cluster's current and desired state.

As an example of holding a cluster's current and desired state, imagine you ask Kubernetes for three Pods of service X (this is the desired state) but Kubernetes notices there are only two Pods up and running for service X (this is the current state). So, Kubernetes will add the missing Pod to bring the state of service X to the desired state, that is, three Pods.

- **Scheduler**: This component is responsible for deciding which worker node the Pods will be created in. It checks worker nodes' available resources, such as CPU and memory, versus what resources are needed for running the Pod. For example, if you have one worker node that has only 1 GB RAM and 1 VCPU left and you need to deploy a Pod that requires 4 GB RAM and 2 VCPUs, then clearly this Pod should not be created in that worker node. So, the Kubernetes scheduler will check other worker nodes and inform the API server about the selected worker nodes for a deployment of that Pod.

- **Controller manager**: Many controller managers run in Kubernetes master nodes, such as the replication controller and namespace controller.

Those controllers run as a `daemon` and periodically check and compare the cluster state via the API server and apply the necessary changes. For example, for the replication controller, we mentioned earlier that the etcd component is responsible for holding (as a data store) the cluster state but the actual component that checks that state is a replication controller. A replication controller asks the API server about the cluster state and the API server will retrieve this information from the etcd data store. The API server is the only component that talks to etcd, while other master node components talk to API servers only.

Now, let's move on to **worker nodes** and see the components that are installed or must be installed on the worker nodes:

- **kubelet**: This is an agent installed in each worker node to communicate with the Kubernetes API server. If there is a Pod that needs to be created in the worker node, the API server will communicate with kubelet about that, then kubelet will do the following:

  A. It will communicate with the container runtime that is installed in the worker node to create or manage the required containers.

  > **Note**
  >
  > A Pod, as we will explain later, is also a container at the end but you can say it is a non-business container.

  B. The container runtime could be Dock containerd, or any other implementation of the Kubernetes **Container Runtime Interface** (**CRI**). The most common container runtime used with Kubernetes is Docker.

  C. kubelet communicates with the container runtime through container runtime-exposed APIs.

  D. It will monitor the Pods/containers running and keep in sync with the API server for any changes that have occurred.

  > **Note**
  >
  > Any container running in the worker node without using kubelet will not be managed by Kubernetes.

- **kube-proxy**: This component is responsible for forwarding traffic to the Pods/containers running inside the worker node. If you think about it, the worker node will have its IP, either a private or public IP. But the Pods/containers running inside this worker node will have IPs that are completely different from the worker node IPs. Kubernetes builds an overlay network layer to manage the networking of Pods inside the worker nodes. This network is Kubernetes' private network, so kube-proxy is similar to a reverse proxy as it forwards traffic from outside or inside the cluster to the right service or application in the worker node.

- **Container runtime**: We covered this before when we talked about kubelet, but in short, it is the container runtime that runs the containers at the end, for example, Docker Engine.

There are other components in Kubernetes clusters, such as **Kubernetes DNS, Ingress controller**, and **the dashboard**; you can refer to the Kubernetes online documentation for more information about these components.

Now, let's talk briefly about the concept of a **Pod** in Kubernetes. A Pod is the main deployment unit in Kubernetes. It is a logical unit, which means there's no actual thing created called a Pod. The actual thing created in the worker node at the end is a container. There is one special container created by kubelet when you create a Pod (more on that container next). This special container is what defines the Pod.

Inside the Pod, you can run one container or many containers. For example, you can run one container of a Spring Boot microservice and another container of a PostgreSQL database in one Pod. However, this is not a recommended pattern. What is recommended is to run one container in one Pod.

Now, to understand more about that special container created when Kubernetes creates a Pod, if you run one container in one Pod and this is the only Pod you've created so far, for example, and then you SSH into the worker node and run `docker ps` to list the Docker containers created (if you use the Docker runtime) in that worker node, then you will find two containers created, not one! You will see your container and another container created called the pause container. This is a special container that we mentioned earlier that creates and handles the concept of a Pod in the worker node.

There is another concept called sidecar containers, which means you run your business application as a single container inside one Pod as recommended. Sidecar containers are additional containers that run in the same Pod and communicate with the business container using localhost (that is, all containers running inside a Pod can talk to each other using localhost). Sidecar containers are usually non-business containers. In other words, they focus on handling cross-cutting concerns, such as security, logging, and monitoring. So, in one Pod, you have the business container focusing on the business side and have one or more side containers in the same Pod communicating with the business container for things such as logging and security.

Let's conclude this section. The following are the basic steps for deploying a containerized application in Kubernetes:

1. If you write containerized software, which can be any software in any programming language, then you just need to add one file to build the software as a container image to your software build files. In Docker, the file is called a `Dockerfile` and it contains all instructions required to build the software into a container image.

2. Now you have a container image stored locally in your build server, so you need to push that container image into a centralized container registry. In Docker, there is a public registry called Docker Hub. You can also push to a private registry.

    A container registry is where your software artifact or container image is stored and ready for installation as a container.

3. (Optional) You can install Docker Engine on a machine (physical or VM), for example. Then, you can run the container image (either from a local image or retrieving the image from the registry where you pushed it) as a running container in that machine. You can do that step in a development environment to test your software, but in production, as explained, you will need a proper container management engine to manage and deploy containers at scale. The next step replaces that step.

4. You need to deploy your container image into a Kubernetes cluster. So, using Kubernetes cluster client tools, such as kubectl or APIs, you send a deployment request (that is, what kind of deployment is required (Pod, Service, and so on) and declare all information required, such as how many Pods should be created and the compute specs of the Pod, in the YAML files) to the Kubernetes API server.

5. Then, the Kubernetes API server will validate the request and coordinate with Kubernetes cluster components, such as etcd and kube-scheduler.

6. The Kubernetes API server talks to kubelet on the selected worker nodes and instructs kubelet to create Pods and other objects in those worker nodes. (Note that the containers created in this step are created by Kubernetes and will be managed by Kubernetes, but in *step 3*, the containers are not created by Kubernetes and will not be managed by Kubernetes accordingly. As we said, *step 3* might be needed for development purposes only – that is, to run a container without Kubernetes.)

7. kubelet talks to the container runtime engine that runs inside the worker nodes to create the required containers. Container images will be retrieved from the container registry. The container image name and container registry information is already defined in the Pod spec YAML file, which is given as input to the Kubernetes API server (see *step 4*).

8. kubelet will keep updating the API server about the status of the running Pods inside those worker nodes.

In the preceding sections, we tried to give a quick overview of Kubernetes. In the next section, we will talk about **different Kubernetes deployment models** and which one to choose for your IoT backend if you choose containers and Kubernetes.

Kubernetes deployment options

We talked about Kubernetes cluster components, that is, master nodes and worker nodes. There are different installation options for Kubernetes clusters, as per the following:

- **Option 1, DIY (in-house):** In this option, you install the clusters with their master nodes and their worker nodes yourself. You are responsible for the master nodes and worker nodes and their internal components that we explained previously.

- **Option 2, managed Kubernetes cluster (in-house):** In this option, you use the services of a specialized Kubernetes vendor who can create, manage, and support your Kubernetes clusters on your behalf.

- **Option 3, managed Kubernetes cluster (on the public cloud):** Most public cloud providers provide Kubernetes managed services where they take care of the master nodes and you take care of the worker nodes. Usually, managing master nodes is a critical and complicated task, hence public cloud providers take care of that part.

  AWS provides **Elastic Kubernetes Service (EKS)**, Microsoft Azure provides **Azure Kubernetes Service (AKS)**, and Google provides **Google Kubernetes Engine (GKE)**.

- **Option 4, serverless Kubernetes cluster (on the public cloud):** Some public cloud providers provide a serverless option for managing containers. In other words, they take care of master nodes and worker nodes and you simply ask to deploy containers and give container images and instructions, such as how many containers you want to run and with what specs.

The question now will be, which option do you use for your IoT backend cloud assuming you build container-based applications and you chose Kubernetes as a container management engine? The answer is, as you might expect by now, it depends on your situation; for example, the first two options in the preceding list might be the only ones available as you can't go for public cloud options.

We recommend option 3, due to the following reasons:

- Mastering and supporting Kubernetes is not an easy task. It requires special skills to operate Kubernetes clusters. By using option 3, the public cloud provider will take care of that difficult task (that is, managing master nodes, which are usually critical and difficult to manage).

- With this option, you still have control of the cluster worker nodes.

- Even if this option is classified as managed services, you are still not coupled with a public cloud provider, as you still follow the standard Kubernetes setup. You just split the responsibility between you and the public cloud provider. If you need to move to another public or private cloud provider, you simply move the container images and not the Kubernetes infrastructure.

So far, we have explored the definition of the cloud and its characteristics, the different cloud models, containers, and Kubernetes to help you as an IoT architect to choose the right infrastructure platform that will host your IoT backend systems and applications.

When it comes to IoT backend systems and solutions, such as the MQTT message broker, streaming engines, databases (SQL and NoSQL), rule engines, and device management, you usually have **two options**: either following the **DIY** model, which includes designing, building, installing, operating, integrating, and testing those IoT systems on your preferred infrastructure option (private, public, VM-based, or container-based), or choosing a **managed IoT platform** that is ready to use and work with immediately.

We recommend using managed IoT platforms for IoT backend systems and we have selected the AWS IoT platform for that purpose. We will cover AWS IoT platform services and solutions in the next few chapters. However, since we are targeting the infrastructure part of the IoT cloud in this chapter, we would like to give a very quick overview of the **AWS cloud** and its services, for two reasons:

- Despite recommending and using the AWS-managed IoT platform, to build a complete end-to-end IoT solution, you might end up using some other non-managed services from AWS cloud services to support your IoT solution. Hence, you should be aware of the AWS cloud in general.

- If you follow the DIY model for building IoT backend systems and solutions and you decide to go with the public cloud option for the infrastructure, then understanding AWS cloud services will help a lot not only for building on the AWS cloud but also for building on other public clouds. Most public cloud providers offer almost the same set of services and solutions. Cloud computing concepts are the same between different public cloud providers; only the names of the services might be different. For example, for compute services of VMs, the AWS service is called **Elastic Compute Cloud** (**EC2**), while Microsoft Azure calls it Azure VMs.

Having discussed the two reasons for its use in this book, in the next section, let's have a quick overview of the AWS cloud.

AWS cloud overview

AWS is the largest and the market-leading public cloud provider. The AWS cloud provides all the different cloud service offerings, such as IaaS, PaaS, SaaS, and CaaS. AWS offers more than 200 services covering **computing, networking, storage, Artificial Intelligence (AI), Machine Learning (ML), game development, security, DevOps, IoT, databases (SQL, NoSQL, graph, and so on), big data, data analytics, Continuous Integration/Continuous Delivery (CI/CD), application development, satellite services, robotics**, and so many other services.

In the following points, let's cover some facts about the AWS cloud:

- AWS global cloud infrastructure is built around the concepts of AWS Regions and Availability Zones.

- An AWS Region is a physical location in the world where AWS has what is called Availability Zones. AWS maintains multiple geographic regions, for example, Regions in North America, South America, Europe, China, Asia Pacific, South Africa, and the Middle East.

  Examples of Regions available in Europe are Frankfurt (`eu-central-1`), Ireland (`eu-west-1`), London (`eu-west-2`), Milan (`eu-south-1`), Paris (`eu-west-3`), and Stockholm (`eu-north-1`).

  At the time of writing this book, AWS has 25 geographic regions around the world with plans to add another 7 regions.

- Within an AWS Region (such as `eu-central-1`) comes the concept of Availability Zones. An Availability Zone is a conceptual or logical concept of a grouping of discrete data centers. Each data center has redundant power, networking, and connectivity and is housed in separate facilities, so one AWS Region has many isolated Availability Zones, and each Availability Zone has many isolated data centers under the hood.

  The Availability Zone name follows the Region name along with adding a zone identifier with letters such as a, b, c, and so on. So, in the `eu-central-1` Region, there are `eu-central-1a` and `eu-central-1b` Availability Zones, for example. If you go deeper behind the scenes, you will discover that the `eu-central-1a` Availability Zone has many data centers supporting it.

So, in the `eu-central-1` Region, which is the Frankfurt Region, in Frankfurt, a couple or more data centers are selected to form the `eu-central-1a` Availability Zone, and they are located some distance from each other to protect them in the event of natural crises, such as floods and earthquakes. This setup gives high availability, durability, and redundancy to all the services launched on the AWS cloud infrastructure.

Each Availability Zone is isolated but the Availability Zones in a Region are connected through low-latency links.

When you launch a resource such as a VM (EC2 in AWS terminology), you can select which Availability Zone(s) you want to launch your resource in or, if not selected, AWS will select a default Availability Zone(s) for that resource.

At the time of writing this book, AWS has about 81 Availability Zones across the globe in different AWS Regions, with a plan to add more.

- AWS also has an edge location presence, which is used mainly for the AWS **Content Delivery Network** (**CDN**) service. At the time of writing this book, AWS has more than 230 edge locations.

- With a need for having ultra-low latency, AWS expands their infrastructure presence or locations to be very close to end users. AWS currently offers the following services in that regard:

  A. **Outposts**: If you have a workload that runs on your data center and you need to have AWS experience in your private data center, then AWS provides the AWS Outposts service, which is AWS hardware that comes with some cloud services running on it. This hardware is deployed in your data center, for example. You can use AWS EC2 services locally in your data center as if they were running on AWS public cloud infrastructure; in reality, they are running on AWS hardware but inside your data center.

  B. **Local Zone**: If your workload is already on AWS and you need to deploy your workload closer to a specific location, Availability Zones might be a bit far in that case. Hence, AWS goes further down the line and provides additional locations that are much closer to end users, called Local Zones.

C. **Wavelength**: This infrastructure service is mainly for covering and supporting ultra-low latency apps. The Wavelength service is deployed on mobile operator edge locations that are very close to the mobile operators' 5G core networks. So, if your mobile, for example, is using a 5G network of operator X, then if you have an app on that mobile, while you are walking in an area covered by mobile operator X, the app will communicate with your backend system, which sits on the mobile operator's X edge location, which might be a few meters away from your location! This is awesome, isn't it? You don't need to worry about the internet traffic latency and multiple network hops required till you reach your backend systems.

Mobile operators – in the case of 5G networks specifically – usually have coverage at a district or even street level. So, by deploying AWS infrastructure services, such as EC2, at that level of coverage, developers can develop and deliver applications that require single-digit millisecond latency, such as game and live video streaming, ML inference at the edge, and **Augmented Reality/Virtual Reality** (**AR/VR**), with the same experience (that is, high availability, performance, and so on) that they have with AWS public cloud infrastructure services.

- Your AWS cloud resources could be deployed inside a **Virtual Private Cloud** (**VPC**). A VPC is your private dedicated network inside the AWS cloud infrastructure. Cloud resources that are usually offered as managed services by AWS can also be deployed outside a VPC at a Region-based level or global-based level and you can access such resources privately from your VPC using AWS PrivateLink (which is an AWS backbone network with no need to traverse over the internet to access such Regions and global services).

A VPC is bound to a Region; in other words, VPC resources can't span different Regions. You can have one or more VPCs in a Region and you can peer (that is, connect) your VPCs inside Regions or across Regions.

- Your relationship with AWS is an AWS account, so you create an AWS account using your email, credit/debit card, and so on. In the AWS account, you can select any Region you want to deploy your workload in and you can deploy your workload in more than one Region to support geo-redundancy or compliance requirements; for example, you can deploy your workload in only one Region to reduce the cost and you depend on the Availability Zones (that is, multi-Availability Zone deployment within one Region) for local-redundancy and high-availability purposes.

You can create as many AWS accounts as you want. Usually, large organizations use AWS. Another service that is also used is called AWS Organizations.

The following diagram shows the kinds of relationships between an AWS account, Region, VPC, different cloud resources, and so on:

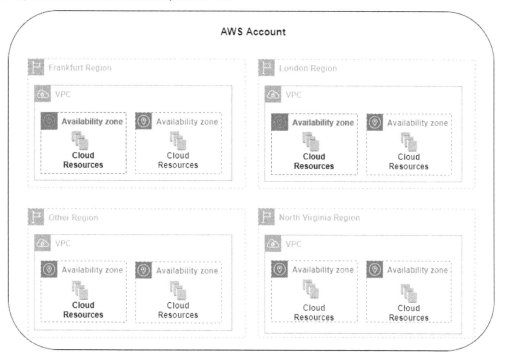

Figure 4.2 – AWS account

This was a very high-level overview of the AWS cloud. In the following points, we will also cover at a very high level the different services you might choose from the 200+ available AWS services for building your backend IoT platform on the AWS cloud:

- **Compute**: For compute, you have different services in AWS, such as the following:

  - **EC2**, or VMs: This is one of the greatest compute services in AWS. It comes under the AWS IaaS offering. AWS provides many EC2 flavors in terms of hardware specs, such as CPU, memory, and network optimization. If you want to deploy applications with a VM-based approach, then EC2 is the best choice. You can use services and features such as autoscaling and Availability Zones to achieve high availability and redundancy. You can also use EC2 in the case of container-based applications by deploying a container management engine such as Kubernetes on top of EC2 machines or even just having a container runtime on those EC2 machines and using bootstrapping and autoscaling features to manage containers running on those EC2 machines without Kubernetes. Or, you can use EKS, which is an AWS-managed container service that uses EC2 for both worker nodes and master nodes.

- **Elastic Beanstalk**: This service comes under the AWS PaaS service offering. It is more abstracted than EC2, as in Elastic Beanstalk you don't care or worry about managing the VM, operating system, and application runtime. For example, if you have a Java application, Elastic Beanstalk gives you a Java runtime environment for running that Java application. You just focus on the code and the Beanstalk service will take care of the deployment, scalability, support, and so on of your application. We don't see this option used in large-scale and production-grade IoT solutions, but it is still a quick and valid option if time and avoiding operational overhead is important to you.

- **Lambda**: This service comes under the AWS serverless offering. If you don't want to worry about maintaining and supporting servers yourself, you can move to the serverless approach. Lambda is the AWS implementation of a serverless architecture paradigm.

- **Outposts**: We have covered this option before. It is used in scenarios where you want to have AWS EC2 machines and other AWS services but in your data center not in AWS cloud data centers. So, you can purchase this service from AWS and AWS will send the hardware and server racks to your data center and they will manage it on your behalf.

- **Container**: We have also touched upon this before, but the following are what AWS provides under this category:

  - **Elastic Container Registry**: This service provides a private container registry where you can upload and manage your container images.

  - **Elastic Container Service**: This is an AWS proprietary container orchestration engine.

  - **EKS**: This is an AWS-managed Kubernetes service. Under this service, AWS provides semi-managed Kubernetes services where AWS takes care of master nodes and you take care of worker nodes. **AWS Fargate** is the other option provided by AWS, which is a fully managed or serverless container service where AWS takes care of everything, that is, master nodes and worker nodes.

- **Storage**: There are so many storage services in AWS but here are the most famous and commonly used ones in IoT solutions:

  - **Amazon Simple Storage Service (S3)**: This is the legendary AWS object storage service that can store any kind of data. As per AWS, S3 supports 11 9s durability. S3 in IoT solutions is usually used as a data lake where all IoT raw data is stored on S3 for further processing and analysis.

- **Amazon Elastic File System** (**EFS**): This service provides managed scalable and elastic filesystem storage options where you can have a shared filesystem between different concurrent clients, such as EC2 machine reads and writes to that shared filesystem.

- **Amazon Elastic Block Store** (**EBS**): This is a block storage device that attaches to EC2 machines. EBS is only accessible by one single EC2 machine at a time and this is what differentiates it from EFS. EBS is usually used as a root device for hosting the operating system of the EC2 machine attached to it, or as an additional/external data device that has its own lifetime outside of the EC2 machine lifetime.

- **Database**: AWS provides a full set of different database options, as follows:

  - **Amazon Relational Database Service** (**RDS**) is a managed SQL database. RDS supports different database engines, such as PostgreSQL, MySQL, MariaDB, Oracle, and Microsoft SQL Server. There is also an AWS proprietary database engine called **Amazon Aurora**. The Aurora engine is compatible with a MySQL and PostgreSQL database engine. We recommend using RDS instead of building, maintaining, and supporting database engines yourself. RDS provides lots of benefits since it is a managed service. For example, you don't need to worry about high availability, data sync between master/slave databases, backup and restore, security, and so many other features of the database. Such features are usually not easy to implement on-premises.

  - **DynamoDB** is a fully managed proprietary NoSQL database service that supports key-value and document data structures. DynamoDB is used heavily in IoT solutions, mainly for building dynamic dashboards and IoT applications.

  - **ElastiCache** is a fully managed in-memory data store and cache service. There are two flavors provided, ElastiCache for Memcached and ElastiCache for Redis.

- **Analytics and big data**: AWS provides a full set of different tools and solutions for building data analytics solutions in the AWS cloud, such as the following:

  Amazon Elastic MapReduce (**EMR**): This is a managed Hadoop cluster in the AWS cloud.

  - **AWS Glue**: This is a fully managed **Extract, Transform, and Load** (**ETL**) service that is used in preparing and loading data into analytics stores for further processing and analysis.

  - **Managed Streaming for Apache Kafka** (**MSK**): This is managed Kafka on AWS.

- **Kinesis**: This is an Amazon proprietary streaming engine. AWS offers four services under the Kinesis family: Kinesis Data Firehose, Kinesis Data Analytics, Kinesis Data Streams, and Kinesis Video Streams.

- **QuickSight** is a managed interactive and **Business Intelligence** (**BI**) dashboard and reporting tool. In QuickSight, you can build different types of charts, reports, and dashboards without any kind of development. Just drag, drop, and configure it in the QuickSight console.

- **AWS Lake Formation**: This is a managed data lake solution in the AWS cloud.

- **Redshift**: This is an AWS proprietary and managed data warehouse solution.

- **IoT**: AWS provides a full set of IoT services and solutions (we will explain these later in the book).

- **ML**: AWS provides a huge set of ML and AI services that enable you to quickly build AI-based apps.

There are so many other services available in AWS, but they are out of the scope of this book.

Summary

In this chapter, you have learned about the IoT cloud or IoT backend, in particular the IoT cloud infrastructure.

Now, you will be able to identify and evaluate the different infrastructure options for your IoT cloud, whether it is private, public, or hybrid cloud, whether you will use a VM or container, whether you will use Kubernetes or not, and which option of Kubernetes to choose: managed Kubernetes services or the DIY Kubernetes option.

You also learned about the AWS cloud, its structure, and the different services and solutions it provides that can help you to design and deliver a production-grade and large-scale end-to-end IoT solution on AWS.

In the next chapter, we will cover the second part of the IoT cloud or IoT backend. We will cover the different IoT applications and solutions provided by the AWS IoT platform.

5
Exploring IoT Platforms

With **IoT platforms**, we come to the most interesting part in designing and building a large-scale IoT solution.

In the previous chapter, we discussed the IoT Cloud (or the IoT backend) infrastructure part. In this chapter, we will go through the software part of the IoT Cloud. In other words, we will explore the different software services and components that should be provided by the IoT Cloud as a backbone for large-scale IoT solutions.

As explained before, in IoT solutions, software or app development spans all of the IoT solution layers. Software is present everywhere in IoT solution layers. In *Chapter 2, The "I" in IoT – IoT Connectivity*, we discussed the different software that could be running on IoT and Edge devices to collect, process, and act upon the collected IoT data through the different IoT sensors attached to IoT devices.

In the IoT backend, we need to understand the different software services deployed in that backend layer and what benefits they provide in terms of designing and building large-scale IoT solutions.

In this chapter, we will cover the IoT platform and the different software services it consists of and provides.

We will not be discussing all IoT platform components and services in this chapter, as some of these components and services will be explained in other, dedicated chapters in the book. Components and solutions such as IoT Device Management and IoT Analytics will each be explained in their dedicated chapters.

The chapter focuses broadly on managed IoT platforms. However, there are some sections in this chapter, covering concepts and technologies such as MQTT, that are common in commercial, managed, open source, and IoT platforms built in-house.

In this chapter, we will cover the following topics:

- IoT platform overview
- IoT platform (buy, build, or use?)
- The AWS IoT platform

IoT platform overview

The IoT platform definition according to Gartner is *an on-premises software suite or a cloud service (IoT platform as a service (PaaS)) that monitors and may manage and control various types of endpoints, often via application business units deployed on the platform.* An IoT platform usually provides (or provisions) web-scale infrastructure capabilities to support basic and advanced IoT solutions and digital business operations.

Practically, we define the IoT platform as a platform that provides and supports the following capabilities for IoT solutions:

- **IoT core**, which covers the IoT Cloud gateway that supports different protocols for receiving and sending IoT data or messages from and to IoT devices – protocols such as MQTT, HTTPS, and MQTT over WebSocket.

- **IoT device management**, which covers how to control IoT devices remotely, firmware updates, and many other device management features.

- **IoT analytics**, which covers how to do analytics on top of the massive amount of IoT data that is generated and collected from IoT devices.

- **IoT edge solutions**, which cover the different software components and services provided at the edge, such as edge applications, analytics, ML and AI, gateways, and protocol adapter services.

- **IoT rule engine**, which covers how to write and manage different rules on the incoming IoT data and how to send such IoT data to the different downstream systems for further processing.

- **IoT storage**, which covers where and how such massive amounts of IoT data will be stored. That storage service is, in fact, a key enabler for some other services, including analytics, ML/AI and IoT dashboards, reporting, and visualization.

- **IoT connectivity management**, which covers the management of different IoT connectivity options that are used in the IoT solution.

- **IoT applications**, which cover the building of different IoT applications, visualizations, and dashboards to enable end users to interact with and control the IoT solution.

- **IoT security**, which covers all security controls to secure the whole IoT solution **end to end** (**E2E**), starting from security controls on IoT devices to security controls on IoT applications and dashboards. Security controls are things such as authentication, authorization, and encryption.

Those are the basic components and solutions that should be provided by the IoT backend platform.

In the next section, we will answer the question, *Should I build an IoT backend platform, buy a ready-made one, or use a managed cloud IoT platform?*

IoT platform (buy, build, or use?)

Before answering the question of whether to buy, build, or use a managed cloud IoT platform, let's agree upon the following facts:

- There is no single vendor or IoT platform that can cover or provide everything needed for building an E2E and robust IoT solution. This fact applies to hyperscale vendors such as Amazon, Microsoft, Google, and others, too.

  The different IoT platform vendors might provide a huge set of components and solution building blocks that assist in building large-scale and production-grade IoT solutions, but you may discover some components that they do not provide or support, for example, IoT connectivity management, in particular, cellular IoT connectivity. For this IoT connectivity management component, you might have to depend on mobile network operators' IoT connectivity platforms for managing the cellular connectivity used in the IoT solution. Another example is IoT device hardware. Most vendors will partner with a specialized device manufacturer partner or vendor to handle the IoT device manufacturing process and so on.

- Even if you choose to build the IoT backend platform yourself, you cannot build everything from scratch as there's no time to do that or point in doing so. Usually, you would start by using an open source ready-made IoT platform and add some additional features that you might require in your IoT project that are not provided out of the box by the open source IoT platform. In other words, you customize an open source IoT platform to fit your needs, or you can build a complete IoT platform from the ground up using some open source components and solutions, and might build some new services that are not available as open source components or solutions.

  For example, if you need an IoT core engine that supports IoT device bi-directional communication using the MQTT protocol, clearly you will not build an MQTT message broker from scratch; you would go with an open source solution such as Mosquitto, RabbitMQ, or HiveMQ to achieve that.

- With the buy option, you have two options. You could buy a commercial IoT platform from a vendor and deploy that commercial platform on your private (or public) cloud. You might need to use this option if you have requirements in terms of deploying the IoT platform on a private cloud. Or, you could buy a cloud-based version of an IoT platform from an IoT platform vendor. With the latter option, you just buy a subscription and the IoT platform is fully deployed and managed by the IoT platform vendor on the IoT platform vendor's cloud (it could be their private cloud or they could use a public cloud).

  Note, we are referring to the use option when we are talking about vendors who offer their IoT platforms from day 1 as a cloud-based IoT platform that is deployed on their cloud and is offered to consumers using a subscription-based billing model.

There are a number of factors you need to consider before deciding on buying, building, or using IoT platforms, to mention a few:

- **Time to market**: Do you want to launch your IoT products quickly and catch some earlier market share or is time not an important factor for you or your company?

  If you need to launch IoT products and solutions quickly, then the buy or use options are the best options for you.

- **Level of customization**: Do you have special technical requirements that drive you away from the existing ready-made IoT platforms as they do not offer the special capabilities that you are looking for, hence must you build one yourself?

If lots of customization is required in many components of a ready-made IoT platform, then the only option would be to build an IoT platform yourself, but if customization is required in only a few components, then you could use a hybrid model, buying a ready-made IoT platform and building some small components that are needed for your special requirements.

- **Cost**: You need to evaluate the cost of building versus buying. Building alone might be cheaper compared to buying, but usually, the total cost of the building option is higher than the cost of the buying option. That is because it is not only about building or implementing an IoT platform, but you also need to cover the cost of ongoing support, maintenance, security, licenses, and so on as well.

  Even if you build your IoT platform using open source components, in production-grade environments, you should use commercial support licenses from external vendors or third parties to support such open source solutions. In production environments, open source doesn't mean free of charge; you still need to support those open source products in production. Clearly, you would not go with community edition open source components in large-scale production environments.

- **Company size**: Usually, start-ups and small business enterprises go with the buy option, while large enterprises with legacy systems and a large number of skilled staff prefer the build option or they buy a commercial IoT platform and deploy it on their private cloud.

Our Recommendation

In general, we prefer and recommend using a cloud-based IoT platform with the flexibility given by the IoT platform vendor to do some customization if needed. Since we mentioned cloud-based platforms, we prefer the vendor to be one of the hyperscale cloud vendors (for example, AWS, Microsoft).

In the next section, let's understand the different types of IoT platforms available on the market nowadays.

IoT platform types

We can classify IoT platforms into the following categories:

- **Generic IoT platforms**: Such IoT platforms provide a generic platform that can be used to deliver any kind of IoT solution. In other words, such platforms focus on providing the technologies or solution building blocks required for building any type of IoT solution.

  Examples: AWS IoT, Microsoft Azure IoT, Google Cloud IoT, IBM Watson IoT, Oracle IoT Cloud, and many others.

- **Specialized IoT platforms**: Such IoT platforms have strength in some specific areas of IoT ecosystems, such as connectivity or device management and industrial IoT. They offer a platform around those areas of expertise and they might offer additional features outside of their area of expertise to enable their platform to compete with other IoT platforms on the market.

  Examples: Verizon ThingSpace, Cisco Kinetic, Software AG Cumulocity IoT, PTC ThingWorx, and so many others.

- **Vertical IoT platforms**: Such IoT platforms provide an IoT platform in a specific business domain or vertical, for example, an IoT smart cities platform, a smart home platform, or an asset tracking platform.

  Some of the generic and specialized IoT platforms provide a ready-made IoT vertical platform to enable developers to build IoT applications in that vertical quickly.

Which IoT platform type should you choose? It depends on your use case; you might just need a vertical IoT platform in the smart city business domain, for example, if that is your company's business area. So you buy that vertical IoT platform and do some customization or configuration if needed and you are good to go for production quickly.

You might choose a specialized IoT platform in device management, for example, if that is what you need for your IoT project, or you might go with a generic IoT platform if you have different IoT use cases and verticals and you need to be able to serve any kind of IoT solutions that you might build now or in the future.

> **Our Recommendation**
>
> We prefer and recommend generic IoT platforms as they give much more flexibility and control when building IoT solutions. Also, they provide you with ready-made IoT verticals that you can use directly without building such verticals yourself. On the AWS IoT platform, for example, you could use a CloudFormation (infrastructure as code) template provided by AWS or its partners for a connected car business case (for example). You would simply run that CloudFormation template, all resources required for that business case would be created automatically in a couple of minutes in your AWS account, and you would be ready to go with a connected car solution. Azure IoT also provides ready-made IoT solutions.

There are lots of generic IoT platforms available on the market. Which one should I choose? We will not mention or recommend a specific one; we will instead give you the **selection criteria** that you can use in selecting the best IoT platform for your IoT solutions.

In the next section, we will go through the list of selection criteria for selecting an IoT platform.

Selection criteria for an IoT platform

On the market today, there are more than 25 IoT platforms of different types (generic, specialized, and vertical-based). In the previous section, we mentioned some names of those IoT platforms.

We believe that lots of choices can be a headache sometimes. In the UK, in the retail sector, there are so many hypermarkets that have lots of different products from different brands; at the same time, there are also supermarkets that follow one (or two, maximum) product brand strategies. Guess what? Many consumers prefer the latter supermarket option. Do you know why? Because of the peace of mind they get. They simply enter the supermarket looking for sugar, for example, they choose it quickly (as there is only one option), and they are done. Easy life.

In the IT world, we believe in the same concept, that is, we need one and only one platform that is very good at doing what we are looking for. We know for sure there's no perfect choice, but a platform that provides 70-80% of what is needed for your use cases can be considered a good choice.

> **Important Note**
>
> As mentioned before, we will not mention or recommend a specific IoT platform or vendor. For simplicity and practicality purposes, we are using the AWS IoT platform in this book as a reference and as an example of an IoT platform, but there's nothing stopping you from using other IoT platforms such as Azure IoT, Google IoT, or IBM IoT. The concepts are almost the same on all those IoT platforms.

The AWS IoT platform is considered one of the best IoT platforms available on the market, so it is good to use it as a reference or example of an IoT platform.

The following are some points that you should consider when selecting an IoT platform:

- **Scalability**: This is a very important point to consider as most IoT projects typically start as a **Proof of concept** (**POC**), and then those POCs move to production if they demonstrate or prove the desirable business value of the business IoT product or solution. At that point, you need your POC to scale accordingly and with future demand as well. So, if you start your POC with an IoT platform with five devices, then you would expect that IoT platform to scale up to cover up to 5 million or even more devices if needed without worrying about changing that IoT platform to another scalable one.

  This is why we recommended IoT platforms from hyperscale cloud providers earlier as they already have the infrastructure required for such a massive IoT scale.

- **Flexibility and modularity**: You need an IoT platform with a flexible, extendable, and modular architecture to help you easily and smoothly customize or add any additional feature you might require in your IoT project. In other words, you need an IoT platform with a plug-and-play architecture style to not get stuck with this IoT platform in handling some mandatory requirements for your IoT solution.

- **Supported feature set**: How many features or services out of the required standard IoT platform features does the candidate IoT platform provide or support? The more the IoT platform provides the core IoT required features plus other additional features or services, the more it becomes a preferable choice for the decision-maker stakeholder.

  Features or services such as IoT Edge, connectivity management, IoT devices (hardware and software), E2E security, IoT application enablement, external integration, or different third parties' ready-made connectors are what differentiate an IoT platform from others.

We recommended earlier IoT platforms coming from hyperscale cloud providers because, besides the IoT platform services they provide, they also provide a massive number of other important services and solutions, such as machine learning, AI, and data analytics. So you will have almost everything you need to build your complete IoT solution or project in one place.

- **Cost**: As usual, you need to consider the cost of the selected IoT platform. Cost usually involves many factors such as the number of devices, the size and frequency of the data ingested from those IoT devices into the IoT platform, and the different features used from the IoT platform offerings.

Each IoT platform provides a pricing model. A pricing model can be subscription-based or based on the number of devices provisioned.

- **Standards and interoperability**: This is a very important point to consider, as you need to check to what level the selected IoT platform follows the well-known IoT industrial standards; for example, does the selected IoT platform support MQTT, HTTPS, and MQTT over WebSocket for IoT Core components or does it support non-standard and proprietary protocols? And for MQTT, is the IoT platform fully compliant with MQTT standard specifications or partially compliant? Does the IoT platform support proper, standard IoT device hardware, IoT communication modules, and so on, or not?

You might say, if we go with a hyperscale cloud IoT platform, then we are stuck or locked in with that hyperscale cloud provider anyway, so it doesn't matter whether their IoT platform supports standard IoT protocols. Well, this is partially true, but think about moving from an IoT platform that follows IoT standards to another IoT platform that follows the same standards. What will the migration or movement look like compared to moving to an IoT platform that doesn't follow well-known IoT standards? Certainly, it (that is, moving from or to a non-standard IoT platform) will not be an easy task, or it might be a truly locked-in scenario where you are stuck with a vendor.

You might decide to go with a hyperscale cloud-based IoT platform even if it is only partially compliant with some standards; for example, AWS IoT and Azure IoT are not 100% compliant with MQTT standard specifications as we will explain in the *Overview of MQTT section* of this chapter. However, your project might not need features that are not covered by AWS IoT, Azure IoT, or any other IoT platform.

- **End user and developer friendliness**: IoT developers are class-A citizens of the IoT platform, so you need to ask whether the selected IoT platform is developer-friendly. In other words, does it provide different programming language SDKs, APIs, documentation, easy-to-develop paradigms, and so on.

  On the other hand, you need to also think about the IoT solution end user who interacts with the applications and solutions that are developed by the IoT developers using that IoT platform. So, you need to check whether the selected IoT platform offers a great user experience to end users or not. Is it easy and quick to onboard or provision an IoT device into that solution? How are engagement and interactive communication managed with the end user? In addition, there are many other factors regarding the user and developer experience with that IoT platform.

- **IoT platform vendor reputation and size**: This is also a very important point to consider. The better the reputation and the greater the size of the IoT platform vendor, the more their IoT platform will have big ecosystems, alliances, and partners supporting and integrating with the IoT platform that give a very good signal of the future of that IoT platform in terms of business continuity in the market, support, the future roadmap, and investment by the IoT platform vendor in that IoT platform.

Those points not only pertain to evaluating managed and cloud-based IoT platforms, but they also apply to evaluating the different open source IoT platforms. If you take the approach of building your own IoT platform, there are many open source IoT platforms on the market, including Kaa, DeviceHive, OpenRemote, ThingsBoard, Thinger, and Zetta.

So far, we have covered the IoT platform definition, features and services, types, and how to choose the best IoT platform for your IoT project or solution.

In the next section, we will explore the AWS IoT platform as a reference for the concept of IoT Cloud-based platforms.

The AWS IoT platform

The AWS Cloud provides lots of different services and solutions across many domains. Let's go through the IoT services provided by AWS:

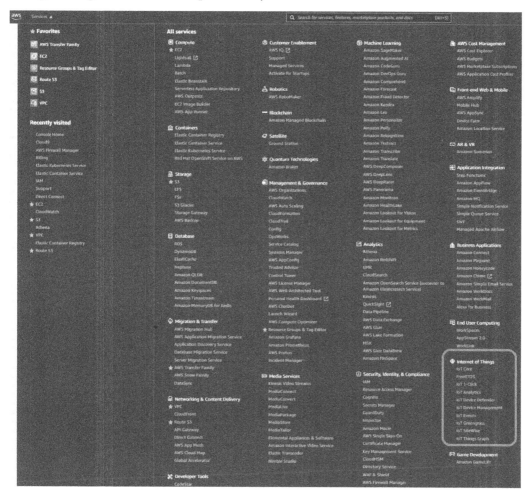

Figure 5.1 – AWS services in the AWS console – Internet of Things and others

In *Figure 5.1*, we have deliberately used a screenshot of all AWS services, and not just those IoT services that are available in the AWS console at the time of writing this book, to stress the fact we mentioned earlier that when you select an IoT platform from one of the hyperscale cloud providers, you also get a huge set of other interesting services that you will need when building a complete E2E IoT solution.

For example, in *Figure 5.1*, you will see about 22 machine learning services and 15 analytic services; under database services, you will see all the different database options that you might need, and the same can be said under storage services, application integration, and so on.

At the time of writing this book, AWS provides 10 services for IoT. The list keeps growing. It started with 2 or 3 services and has now become 10. We expect the list to be added to in the future with additional IoT services and solutions.

Now let's go through the list of those AWS IoT services and see a roadmap of how we will cover those services in this book:

- **IoT Core**: As the name suggests, this is the core AWS service for IoT workloads. Basically, it is the IoT gateway that handles bi-directional communication with IoT devices.

- **FreeRTOS**: We already covered this service in *Chapter 3, The "T" in IoT – Devices and the Edge*.

- **AWS IoT 1-Click**: We will briefly explain this service in this chapter, in the *Other AWS IoT services section*.

- **IoT Analytics**: This service will be covered in *Chapter 7, In the End, It Is All about Data, Isn't It?*.

- **IoT Device Management**: This service will be covered in *Chapter 6, Understanding IoT Device Management*.

- **IoT Device Defender**: This service will be covered in *Chapter 6, Understanding IoT Device Management*.

- **IoT Events**: This service will be covered in *Chapter 7, In the End, It Is All about Data, Isn't It?*.

- **IoT Greengrass**: We already covered this service in *Chapter 3, The "T" in IoT – Devices and the Edge*.

- **IoT SiteWise**: This service will be covered in *Chapter 7, In the End, It Is All about "Data, Isn't It?*.

- **IoT Things Graph**: We will briefly explain this service in this chapter, in the *Other AWS IoT services section*.

Before we deep dive into AWS IoT Core in the next section, let's discuss the MQTT protocol, the most well-known and widely used IoT application communication protocol.

Any IoT Core component, whether you build it yourself or use one of the managed IoT Cloud platforms, must have an MQTT message broker component, so it is good to know about the MQTT concepts before we explain the AWS IoT Core component.

Overview of MQTT

Here's a quote from the MQTT official website (`https://mqtt.org/`):

MQTT is an OASIS standard messaging protocol for the Internet of Things (IoT). It is designed as an extremely lightweight publish/subscribe messaging transport that is ideal for connecting remote devices with a small code footprint and minimal network bandwidth. MQTT today is used in a wide variety of industries, such as automotive, manufacturing, telecommunications, oil, and gas.

The MQTT protocol is not new. It was invented in 1999 to handle a number of technical challenges and scenarios such as sensing and acting on oil pipelines that are deployed in remote areas where there are no traditional or consistent connectivity options available. The satellite connectivity option that is used in such cases is usually very expensive and requires a special communication protocol that is simple, lightweight, bandwidth-efficient, handles the quality of service for message delivery, and is data -agnostic and session -aware.

The MQTT protocol was the answer to those challenges as it has the following features:

- **Standard and open**: MQTT is an OASIS standard. The specification is managed by the OASIS MQTT Technical Committee.

> **Note**
> There are broadly two specifications for MQTT, MQTT 3.1 (and 3.1.1) and the most recent one, which is MQTT 5. MQTT 3.1 (or 3.1.1) is the most widely used MQTT protocol on the market as of now given that MQTT 5 was only introduced fairly recently, that is, 2019.

- **Built on top of the TCP/IP protocol**: The MQTT protocol is used in the application layer (layer 7), while the TCP protocol is used in the transmission layer (layer 4). The MQTT protocol has been designed to support reliable communication over unreliable and unpredictable communication channels such as the internet.

> **Note**
> MQTT uses or requires TCP as a transport protocol as TCP is a much more reliable, controlled (that is, it handles the order of messages, error checks, and so on in the transport layer), and session-oriented transport protocol. Those TCP characteristics or features are required by the MQTT protocol to support communication reliability.

MQTT also provides a heartbeat mechanism to detect any loss in TCP connections between MQTT clients and the MQTT broker/server and act accordingly. Hence, MQTT is not only dependent on the TCP protocol when it comes to reliability; it goes even further by using additional mechanisms such as Heartbeat for more reliability and consistency between MQTT clients and MQTT brokers.

We mentioned before that IoT endpoint devices are usually low-power, low-cost, and resource-limited devices. IoT endpoint devices with such constrained and limited characteristics bring new technologies to many domains. For example, in the device domain, lots of technologies and improvements have been introduced around microcontrollers in terms of their size, processing capacity, and so on; in the connectivity domain, technologies such as **Low-Power Wide-Area Networks (LPWANs)** have been introduced.

Those constrained or limited device characteristics also bring new technologies and protocols to the IoT application communication protocol, including **MQTT for Sensor Networks(MQTT-SN)**, which uses UDP for the transport layer instead of TCP and was introduced to cover such constrained or limited device requirements.

We will not explain MQTT-SN in this section as traditional MQTT, which runs on top of TCP, is the most widely used IoT application protocol in lots of production-grade IoT solutions. Also, the differences between MQTT-SN and classic MQTT are not big, so understanding MQTT will accelerate the understanding of MQTT-SN if needed.

The reason why we will only focus here on classic MQTT, which is run over TCP/IP, is that earlier in the book, when we discussed the deployment pattern of IoT solutions, we mentioned that the most common and frequently used deployment pattern is one that is low-power, low-cost, and has limited endpoint IoT devices in the field connected to a gateway or edge device, which is somehow powerful compared to the IoT endpoint devices. IoT endpoint devices can run any protocol (IP or non-IP) to communicate with an IoT gateway device. The IoT gateway or edge device usually has internet connectivity and has to run the TCP/IP software stack, meaning the gateway or edge device will be able to communicate with the cloud-based IoT Core component using MQTT, which runs on top of TCP/IP.

Locally, that is, between the IoT endpoint device and IoT gateway device, you can use MQTT-SN, and between the IoT gateway and the IoT Cloud, you can use MQTT over TCP/IP.

One other thing: most IoT core engines currently run MQTT, not MQTT-SN.

- **Binary**: This is not like HTTP or other text-based protocols; it is designed to be a super-efficient and lightweight protocol, hence it uses binary format. The minimum packet size of an MQTT message is only 2 bytes, so that makes MQTT super-efficient on wires when transferred between MQTT clients and MQTT brokers.

- **Bi-directional**: You can send an MQTT message from a device to an MQTT message broker or from an MQTT message broker to a device.

- **Data-agnostic**: You can send a message or payload in whatever format you want, for example, XML or JSON.

- **Built for push communication**: The MQTT message broker will send/push a message to all subscribers. This pattern reduces the latency of sending/receiving messages between devices.

- **Built for constrained devices**: MQTT software libraries require a low footprint in memory and processing, hence the MQTT protocol is the best IoT protocol in the application layer as most IoT devices are very constrained and limit devices in terms of memory and processing, as explained previously.

- **Scalable**: The pub-sub or the publish-subscribe architecture paradigm that the MQTT protocol is built upon is a scalable communication architecture paradigm. We will explain that architecture paradigm later in this chapter.

- **Secure**: It runs on **Transport Layer Security** (**TLS**) and it provides an authentication and authorization mechanism as part of the protocol specifications. We will explain that later in this chapter.

In the next section, let's understand the MQTT pub-sub architecture and how it works.

MQTT pub-sub architecture

Now, let's talk about the MQTT pub-sub architecture model and explain its different components:

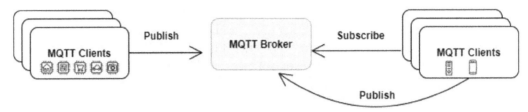

Figure 5.2 – MQTT pub-sub architecture

We have the following components in the typical MQTT pub-sub architecture:

- **MQTT broker**: This is the MQTT server that receives, buffers, and manages the MQTT messages coming from MQTT clients (who act as publishers) and then it filters and delivers the messages to the interested recipient MQTT clients (who act as subscribers).

  The MQTT clients do not talk to each other directly; they typically have to communicate with the central system (MQTT broker) that enables communication between them.

  There are many MQTT message brokers, both open source and commercial, available on the market, including Mosquitto, HiveMQ, and VerneMQ. There are also cloud-based, managed MQTT brokers such as AWS IoT Core and Azure IoT Core.

  > **Note**
  >
  > Some MQTT message brokers might not be fully compliant with the MQTT standard specifications. In AWS IoT Core, Microsoft Azure IoT Hub, and so on, there are some limitations and deviations from MQTT standard specifications, as we will explain later in this section.

- **MQTT clients**: MQTT clients can play the role of MQTT publishers, which could be any system or device that can publish an MQTT message to the MQTT broker topics. Also, MQTT clients can play the role of MQTT subscribers, which could be any system or device that can receive MQTT messages from the MQTT broker through subscription to MQTT topics.

On your system or device that acts as an MQTT client, you can implement the MQTT protocol yourself from scratch in whatever programming language you use on that device. In the end, you need to construct and send an MQTT packet over the TCP layer as per the MQTT protocol specification, but usually, this is not the case (that is, you don't need to build an MQTT client library from scratch) as there is already a big set of software libraries and SDKs supporting the MQTT protocol in different languages, such as Java, C, C++, Python, and C#. The most well-known open source MQTT library is Eclipse Paho.

When we talk about AWS IoT Core, we should mention that AWS provides IoT device SDKs. Those SDKs implement the MQTT protocol under the hood as per the AWS IoT Core MQTT broker implementation. Those SDKs enable the IoT devices to connect (that is, act as an MQTT publisher) or subscribe (that is, act as MQTT subscribers) to AWS IoT Core, which is the AWS MQTT broker implementation.

> **Note**
> If you have an IoT device that wants to communicate with AWS IoT Core, for example, then you have to use AWS-provided SDKs or libraries on your IoT device. In other words, the standard MQTT client library might not work with AWS IoT Core. This is related to the small deviation we referred to earlier regarding AWS IoT Core, Azure IoT Hub, and so on, from MQTT standard specifications.

The publish-subscribe pattern is scalable and much better than the client-server (sync communication) pattern. For scalability, you can add additional publishers or subscribers without impacting the current publishers and subscribers.

In the pub-sub pattern, the publishers and subscribers are fully decoupled. The publishers and subscribers don't need to know about each other and do not need to be online at the same time in order to communicate. The message broker buffers and handles the MQTT messages until such time as the subscribers come online again.

The only challenge or single point of failure in that pattern is the message broker, since, if the message broker fails, the communication link will be broken between MQTT clients. To overcome that challenge, most MQTT message broker providers support cluster-based MQTT message brokers, that is, more than one MQTT message broker node running in a cluster node to avoid a single point of failure.

In the next section, let's look at some MQTT concepts that will help with understanding how the MQTT protocol works.

MQTT concepts

On an IoT device, publishing an MQTT message to the MQTT message broker is a super-easy task. As we mentioned earlier, you just need to download – in whatever programming language you use – and install the MQTT software client libraries or SDKs. Then, in your source code, you reference or call those MQTT client libraries' APIs to publish, subscribe, or unsubscribe to an MQTT message broker. In other words, you just need to call publish, subscribe, or unsubscribe methods or APIs provided by those ready-made MQTT client libraries or SDKs, and you are done.

Under the hood, those MQTT client APIs or methods implement the MQTT protocol specification. So, let's see how the protocol works to help you evaluate the selected IoT platform and check whether you should go with that IoT platform implementation of the MQTT protocol specifications.

Example – MQTT SDK

If you are a Java developer (for example) and you have software (written in Java) that is running on an IoT device and you would like to connect or communicate with the AWS IoT Core service over MQTT or MQTT over the WebSocket protocol, then you can use the AWS IoT Device SDK for Java, which is available here: `https://github.com/aws/aws-iot-device-sdk-java`. This SDK not only provides support for MQTT communication; it also supports other features, such as integration with the AWS IoT Device Shadow service, which we will talk about later. The AWS IoT Device SDK is available in other programming languages, including C++, Python, and JavaScript.

Before explaining how MQTT clients interact and communicate with the MQTT server, let's understand the following concepts in MQTT first:

- **Topics**: Topics are the destination, the communication medium, or the storage that publishers and subscribers interact through. In simple form, topics are just an UTF8-string that a broker uses to filter messages and deliver them to the right connected subscribers.

A topic name consists of one or more levels. Each topic level is separated by a forward slash (/). In other words, the topic name represents a logical hierarchy or tree similar to the filesystem folders and file tree. For example, you might have a topic name called `city1/district1/park1/floor1/lotlot1`. It is only one string, but it represents multiple levels, such as city level, district level, and park level. So, by choosing to name the topic as `city1/district1/park1/floor1/lotlot1`, you are referring to parking lot number 1, which exists on floor number 1, which exists in parking number 1, which exists in a district called district 1, which exists in a city called city 1.

Topic names are case-sensitive. The length of the topic name can be from one character long to more than 65,000, but it is preferable to keep the name of the topic short. The shorter the topic name is, the more bandwidth you save on the wire.

MQTT 5 introduced a new feature called Topic Alias to enable MQTT clients to publish or subscribe to topics by using numbers or integer values instead of strings. The MQTT message broker on the other side (the server or cloud side) will do the mapping of those aliases or numbers to the proper topic names. This feature helps in reducing bandwidth as the numbers or integers will be smaller than strings in size. Accordingly, the MQTT packet size will be reduced, meaning it will require a small bandwidth to be transferred to the MQTT message broker.

Topics are created dynamically. This is a very important point to remember as you don't need to create topics before publishing; you can just create the topic, publish it, and subscribe to it dynamically at runtime. You can create thousands or millions of topics depending on the message broker's capacity.

It is also important to understand the use of wildcards when subscribing to topics. Here are the rules for wildcards:

- +: This is a single-level wildcard. For example, you might be interested in subscribing to all parking floors in `park1` to get sensor data of parking `lot1` on those floors. So, instead of subscribing to each floor separately, you can subscribe to `city1/district1/park1/+/lot1`. This will save resources and offer improved efficiency for message brokers and MQTT clients.

- #: This is a multi-level wildcard. In the example we have mentioned, let's say you need to subscribe to all parks with their floors and get all parking lots' (not only `lot1`) sensor data in `district1`. So, in this case, you just need to subscribe to `city1/district1/#`.

Here are some of the best practices you should use when dealing with MQTT topics:

I. For topic names, don't use spaces, don't start a topic name with a forward slash, and use only ASCII characters.

II. Make the topic name as short as you can. This is important to save some bandwidth.

III. Use the unique identifier of the MQTT client as part of the topic name. This helps the message broker in filtering and delivering messages to the right subscribers.

IV. Don't subscribe to root #. This might kill your subscriber clients since, by subscribing to root #, you subscribe to receive all messages sent to all message broker topics. Your client might not be able to cope with that traffic, be overloaded, and then crash in the end. Some message brokers do not allow subscribing to root #.

- **QoS**: When it comes to MQTT and the message delivery guarantee between the communication participants (the client and broker or server), there are three different levels or three QoS levels, as per the following:

  - **QoS 0**: On this level, at most one message will be delivered, which means it is a fire-and -forget mode. In other words, the publisher will send the message and will not wait for an acknowledgment packet from the message broker that confirms receipt of the MQTT message. The message could be lost on that QoS level and no one would be aware of it.

  - **QoS 1**: On this level, at least one message will be delivered, so when the publisher publishes an MQTT message to the message broker, the publisher will receive from the message broker a PUBACK packet to acknowledge receipt of the message.

  - **QoS 2**: On this level, exactly one message will be delivered, so the publisher will send or publish an MQTT message with QoS 2 to the broker, and then the broker will respond with the PUBREC packet. The publisher will then reply with PUBREL (this packet is a response to the PUBREC packet). Finally, the broker will send PUBCOMP to complete the flow to make sure exactly one message has been delivered. So, even if there's a drop in the network during communication, only one message will be delivered in the end as the following has to be completed between MQTT clients and brokers.

Here are some of the best practices you should use when dealing with QoS or when you want to identify which one of the QoS levels to use:

- Use **QoS 0** if your IoT business case can afford to lose some messages. For example, losing some room temperature degree messages that your IoT device sends to your IoT Cloud every few seconds is not a problem.

- You can also use **QoS 0** if you have a limited network bandwidth as QoS 0 requires less overhead. It uses the fire -and -forget communication mode.

- Use **QoS 1** as the default QoS option. This option offers the best trade-off between **QoS 0** and **QoS 1** as it guarantees the delivery of messages from one end and it uses less bandwidth compared to QoS 2.

- Use **QoS 2** when you are not limited in terms of network bandwidth so the communication protocol overhead (that is, acknowledgment packets, and so on) is acceptable and low performance or high latency is acceptable as well.

It is worth mentioning that most message brokers use QoS 1 as the default option.

The QoS feature is one of the features that makes AWS IoT Core, Azure IoT Hub, and so on deviate a bit from, or not be fully compliant with, the MQTT standard specification. AWS IoT Core and Azure IoT Hub do not support QoS 2, for example.

Also, AWS IoT Core has a number of limitations in terms of its features, such as the QoS 1 retry interval, which is only 1 hour in the case of AWS, meaning that if the device was offline for more than 1 hour, the message would not be delivered. In some IoT projects or use cases, that might not be acceptable, and you must find a solution for that, such as a supporting QoS 2.

- **Last Will and Testament**: This is a unique feature of the MQTT protocol. The idea is simple: when the MQTT clients connect to the MQTT message broker, they send some information about their last will message, last will topic, and so on. If the MQTT clients disconnect ungracefully, the MQTT message broker will send a notification to all subscribers on behalf of those MQTT clients about the last will of those disconnected MQTT clients.

The broker will send the last will and testament message when the following occurs:

- The broker detects I/O errors or network errors.

- The MQTT client doesn't communicate with the message broker for a period more than the keep-alive configured period. We will explain the keep-alive concept later.

- The MQTT client doesn't send a `DISCONNECT` packet before it closes the network connection.

- The broker closes the connection due to whatever error occurred on the broker side.

You can leverage the concept of last will in your IoT solution design. How? You can use it for error handling scenarios. For example, you could subscribe to the last will topic and if you receive messages on it, this means the IoT device is disconnected.

Then, you can send an SMS or email or call an API to notify the user of that error.

- **Persistent sessions and message queuing**: This feature enables the MQTT message broker to remember the clients connected to it so that if those clients disconnect and reconnect again, the broker will restore their session instead of creating new sessions, connections, and so on for them.

The broker stores or queues messages of QoS 1 and QoS 2 that the MQTT clients miss when they are offline, so once the MQTT clients connect again, they will get the queued messages of QoS 1 and QoS 2. The broker also stores unacknowledged QoS messages.

MQTT clients define in the MQTT connection packet whether they need the session to be persistent or cleared. If a cleared session is configured, then no messages or sessions will be persisted.

By default, messages stored in a broker are not expired but some brokers implement a number of expiration techniques to expire stored messages based on some threshold. It is important that this expiration technique is implemented to avoid DDoS attacks on the MQTT broker as MQTT brokers might be flooded with DDoS messages. Without expiration, the MQTT broker would crash eventually.

- **Retained messages**: This is also a unique feature of the MQTT protocol. A retained message is the last known good value persisted by the MQTT message broker for a topic. When new subscribers subscribe to a topic, if retained messages are enabled in that topic, the new subscribers do not need to wait for incoming messages from the publishers, which might take some time. With the retained message feature enabled on that topic, they will immediately get the retained messages of that topic once they subscribe.

The publishers set the flag of a retained message in the `publish` packet. The flag could be true, meaning to persist the retained message, or it could be false, meaning there's no need to store a retained message for that topic. Only one retained message is stored per topic. Every time the publishers set the flag of a retained message to true, that message will override the previous one, and so on.

- **Keep alive**: The MQTT protocol uses TCP as the transport layer, which is a reliable transmission protocol, that is, if a client disconnects, the TCP session will be closed and the broker will detect the disconnection. However, there are some scenarios where special handling of a TCP connection is required. One of those scenarios is what is known as a half-open TCP connection, where one entity (the MQTT client, for example) knows or believes it is disconnected from the TCP session that it has with the MQTT broker, while the other side (the MQTT broker, for example) knows or believes the TCP session they have with the client is still alive or active.

  To solve that challenge, you need to handle it in the application layer, not in the transport layer, by using protocols such as the **Internet Control Message Protocol (ICMP)**/PING or a technique such as a keep-alive technique between the client and the server (or the broker).

  The way keep-alive works in MQTT is that the publisher client will send the keep-alive value in the `publish` packet. Let's say the value is set to 60 seconds. This means that the broker will check whether any requests are coming from the client in those 60 seconds. If there are no requests, then the broker will assume that the client is disconnected.

  Usually, the client implements keep-alive by repeatedly sending the PINGREQ MQTT packet (like PING), and then the broker will respond with the PINGRESP MQTT packet. This is done under the hood, and it is executed by MQTT software client libraries/SDKs that you use in your project.

  The keep-alive default value is 60 seconds, but it can be increased up to 18 hours or it can be disabled (that is, no keep-alive, but this is not recommended).

  There is another feature related to the keep-alive feature, called **client takeover**, which means, in the scenario of **a half TCP open connection**, the client disconnects but the broker is still connected. The client reconnects with a new TCP session with the broker. The broker will then have two sessions open with the same client. Client takeover means the second TCP session will continue and the first one, that is, the half-open one, will be killed by the broker. This feature is important where clients are stuck in communication and they have to reconnect again to get out of that problem.

Those are the basic concepts you should understand about the MQTT protocol at a high level.

In the keep-alive concept, we explained the client takeover feature, which is a very good feature, but, at the same time, it triggers another point to consider in relation to MQTT, which is **MQTT security** or authentication and authorization in MQTT. With the client takeover feature, it is important to authenticate MQTT clients with the broker to not have unauthorized and unauthenticated MQTT clients take over MQTT sessions of legitimate and authorized MQTT clients, meaning security in MQTT is very important. Let's see how security works in MQTT in the next section.

MQTT security

Security in the MQTT protocol, or in general, goes through different layers of the application stack. Each layer prevents a different kind of attack, so for the MQTT application stack, we have the following:

- **The network layer**: In this layer, security controls such as VPN/IPSec or a dedicated private link between MQTT clients and MQTT brokers can prevent security attacks on the network layer.

- **The transport layer**: In this layer, there are security controls such as the use of TLS/SSL, which is used to encrypt the traffic between MQTT clients and MQTT brokers. Using TLS/SSL, you can support one-way SSL (that is, traffic is encrypted between MQTT clients and MQTT brokers and clients are 100% sure they are talking to the right MQTT broker) or you can support two-way SSL or mutual-SSL (that is, similar to one-way SSL with the addition that the MQTT broker is also 100% sure it is talking to the right MQTT client).

  You can encrypt the payload of an MQTT message in the application layer instead of using TLS/SSL in the transport layer if you are worried about the overload of TLS/SSL, but usually, it is a security recommendation to enable TLS/SSL in the transport layer.

- **The application layer**: The MQTT protocol standard specification provides a client identifier and username/password credentials to authenticate MQTT clients connecting to the MQTT broker. This is for authentication. For authorization, it is up to the broker's implementation how it will do authorization, so each broker provides their own implementation of authentication and authorization.

For authentication, in MQTT, we have the following authentication methods:

- **Username/password**: These are optional in the protocol standards. It is allowed to send a username without a password, but it is not allowed to send a password without a username.

- **Client ID**: Each MQTT client must have a unique client ID. That client ID can be used as a way of authentication. Usually, it is used as an additional authentication step with a username/password, but there's nothing to stop you from using a client ID to authenticate MQTT clients.

- **Client certificate**: This is a very interesting authentication method and it is commonly used in production-grade IoT solutions.

  The authentication of MQTT clients using a client certificate will be done at the transport level, that is, during the TLS handshake and before the MQTT client connects to the MQTT broker. This will help in saving the MQTT broker's resources.

There are different ways of implementing authorization in the MQTT broker. The common ones are the following:

- **Access Control List (ACL)**: Here, the MQTT broker can associate the ACL with a resource, that is, a topic. The ACL contains permissions of who (the principle) can access that resource and which operations, for example, read, write, and execute, are allowed for that principle.

- **Role-based access control**: Here, the MQTT broker associates a role with a resource, that is, a topic. The role abstracts the permissions of what the user (principle) holding that role can do with this resource and which operations, for example, read, write, and execute, are allowed for that principle on that resource.

There are other authorization mechanisms, such as fine-grained authorization, but anyway, the bottom line here is that each MQTT broker can have their own implementation of authentication and authorization, so you need to follow the documentation and best practices offered by your selected MQTT broker. When we talk about AWS IoT Core, you will see how AWS handles authentication and authorization in the AWS IoT Core message broker.

After understanding the MQTT architecture, basic concepts, and MQTT security, let's see in the next section how the MQTT client interacts with the MQTT broker for publishing messages. In other words, let's see how the MQTT connection flow works.

Note, with MQTT software client libraries/SDKs, you don't need to understand the following section as the software client libraries will take care of that and will abstract the internals of the connect, publish, subscribe, and unsubscribe APIs for you, but we explain such details just in case you use a programming language that has no ready-made MQTT software client libraries/SDKs available for that language yet and you have to implement the MQTT client yourself in that programming language. The next section covers the connection part only, and does not cover publish, subscribe, and unsubscribe internal APIs.

If you are implementing the MQTT software client yourself, then the best place to check such details would be the MQTT protocol standard specification.

MQTT connection flow

In MQTT, we have two parties involved in communication: the MQTT clients and the MQTT server or broker. The **connection flow** between them is as follows:

1. For the MQTT clients to connect to the MQTT server, they establish a TCP connection or TLS connection. If an encryption channel is required with the MQTT server, in that step, the MQTT client sends a connect packet to the MQTT server. The MQTT server will authenticate the MQTT clients, check the authorization policy, that is, what is allowed/not allowed for those MQTT clients, and then the MQTT server will reply to that connect packet with a connack packet.

    The MQTT connect packet contains the following information:

    - **clientID**: This field holds the unique identification for the MQTT client. It could be the IoT device serial number, the IoT device MAC address, or anything that makes it unique.

    - **Username**: This field is used with the following password field for authentication purposes. This field is optional, so it can be empty.

    - **Password**: This is the password related to the username field. This field is optional but if it is populated, then the username field must be populated as well.

    - **CleanSession**: This is a Boolean field. If it is set to true, it means if the MQTT client disconnected for whatever reason, then the current session between the client and broker is cleared and not persisted. If it is set to false, it means the session between the client and broker will be persisted, so when the client reconnects again, the old session with all the information persisted will be retrieved and the client will continue the session. This is helpful in the case of constrained IoT devices as you don't need to redo operations such as resubscribe.

Here is what is stored for a persisted session on the message broker side. The broker uses `clientId` to retrieve the client's persisted session:

- All the subscriptions of the client, that is, the topics the client subscribed to and so on

- All messages from QoS 1 or 2 that have not yet been confirmed by the client

- All messages from QoS 1 or 2 that the client missed while it was offline

- All QoS 2 messages that are not completely acknowledged:

  i. **LastWillTopic**: This field holds the topic name where the `LastWill` message will be sent.

  ii. **LastWillQoS**: This field holds the QoS of the `LastWill` message.

  iii. **LastWillMessage**: This field holds the actual `LastWill` message sent to the MQTT broker.

  iv. **LastWillRetain**: This is a flag to enable or disable the retention of the `LastWill` message.

  v. **KeepAlive**: This is a time interval in seconds for keeping the connection between the client and the broker alive.

After the broker performs the authentication and authorization of the client request, it replies to the client with a `Connack` packet that holds the following information:

- **SessionPresent**: This is a Boolean value of true or false. The true value means that the broker is saying to the client that it already has a previous session for that client. If the client connects with the `clearSession` flag set to `true`, then this `sessionpresent` flag will always return as false.

- **ReturnCode**: This code tells the client whether the connection attempt was successful. The `0` code means the connection was accepted, `1` means the connection was refused – *unacceptable protocol version*, `2` means the connection was refused, the identifier was rejected, and so on.

Now that we have covered MQTT in some detail, in the next section, let's talk about AWS IoT Core and its different components.

AWS IoT Core

AWS IoT Core is a fully managed, secure, and scalable service that supports bi-directional communication between internet-connected IoT devices and the AWS Cloud.

AWS IoT Core is the receiver system for the incoming IoT telemetry data that is generated by IoT devices and sent across the internet. It also enables users and applications to control and communicate with IoT devices.

AWS IoT Core consists of many components or solution building blocks:

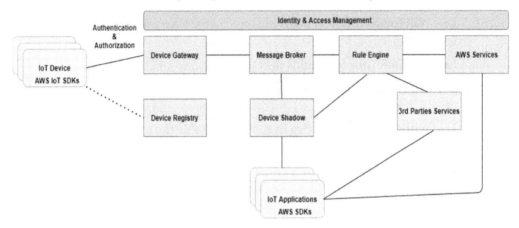

Figure 5.3 – AWS IoT Core components

Let's explain the IoT Core components in detail:

- **AWS IoT SDKs**: Starting from the IoT devices, as explained in the previous chapters, you have an IoT device that has a real-time operating system such as FreeRTOS, for example, or even doesn't have an operating system at all. You need to write embedded software to interact with sensors/actuators connected to this device and then send such readings (that is, IoT data) to the MQTT or HTTP message broker that is running on the IoT Cloud (or IoT Edge cloud) for further processing.

    So, instead of writing software to connect to the MQTT message broker that runs on the AWS Cloud (that is, AWS IoT Core) or the AWS edge cloud (that is, the MQTT broker on Greengrass), AWS provides a set of SDKs to help you quickly and smoothly build applications or software on those IoT devices for sending and receiving messages from AWS IoT Core.

If you use Amazon FreeRTOS as an operating system for your IoT devices, then AWS IoT SDKs and software client libraries will come pre-installed with the Amazon FreeRTOS operating system, or you can choose to add such add-ons or libraries to the FreeRTOS version when downloading and installing it. Otherwise, if you use something else, then you can simply download and install those AWS IoT SDKs with the preferred programming language you want, given that AWS provides SDKs in that programming language.

There are many SDKs provided by AWS that support AWS IoT services, such as AWS IoT Device SDKs, mobile SDKs, and AWS IoT Device Client.

AWS supports and provides SDKs in many different programming languages, including C++, Python, JavaScript, and Java. It also supports and provides AWS IoT APIs with AWS IoT SDKs built on top of them. So, if you use other programming languages that have no SDKs provided by AWS, you can still integrate with AWS IoT Core using the AWS IoT APIs, which are REST (HTTPS) APIs.

To conclude in relation to this component, you don't need to worry about the software code needed for connecting, authenticating, publishing, and receiving IoT messages to AWS IoT Core; AWS IoT SDKs will take care of that part on your behalf.

- **Device registry**: AWS provides a database or a registry that helps in managing things or devices. A thing in the AWS device registry is a logical representation of an actual or physical IoT device or sensor. The device registry holds information or metadata about IoT devices that are represented by an object (thing) in the device registry.

  Information or metadata of IoT devices could be something such as device manufacturer, make, model, serial number, or any information you would like to store in relation to an IoT device in AWS IoT Core.

The device registry also establishes the identity of IoT devices; for example, it stores what kind of authentication methods an IoT device uses, stores a unique ID of the IoT device, and so on.

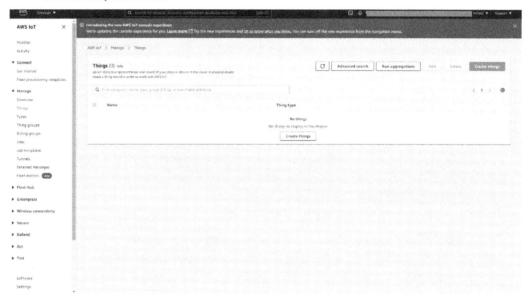

Figure 5.4 – AWS IoT Core console

There are many objects stored in the device registry, such as things, thing types, and thing groups.

A thing is the representation of a single IoT device. You can create a thing directly, but it is recommended to create a thing type object first that holds properties of IoT devices that share some common properties, and then you create different things from those thing types. For example, you create a thing type for Smart meter things and another thing type for Smart parking things. The idea is similar to the concepts of class (thing type) and object (thing type) in object-oriented programming.

A thing group, as the name suggests, is for grouping and organizing things to simplify some device management operations or tasks such as finding a specific device or applying batching to a specific device.

There are other objects, as shown in *Figure 5.3*. In this section, and in this chapter in general, we just need you to be aware of the thing object as this is the main object in AWS IoT Core that virtually represents the actual IoT device.

- **IoT communication protocols**: AWS IoT Core supports the following protocols:

  - MQTT

  - HTTPS

  - MQTT over the WebSocket protocol

  MQTT is the preferred protocol for IoT devices since it requires a low footprint in terms of memory and processing. MQTT is lighter than the HTTP(S) protocol; however, AWS IoT Core offers an HTTPS endpoint for IoT devices that support and can communicate with AWS IoT Core using the HTTPS protocol.

  From a security perspective, usually ports 80(HTTP) and 443(HTTPS) are okay to be open for inbound and outbound traffic, but other ports, such as the MQTT port, are usually not allowed. If you still need to use the MQTT protocol while such a security policy is in place, then you can go with the third protocol option offered by AWS IoT Core, which is MQTT over WebSocket.

  MQTT over WebSocket is also used to support browser-based applications, that is, if you need to send or receive MQTT messages in a browser-based application.

- **Authentication and authorization**: Before IoT devices start sending or receiving messages to or from AWS IoT Core, you need to authenticate those IoT devices. AWS IoT Core will need to make sure those IoT devices truly belong to you.

  Not only is the authentication of those IoT devices needed, but you also need to make sure those IoT devices are authorized to do what they are trying to do. For example, you might have an authorization policy to allow specific IoT devices to publish or subscribe to a specific topic, and so on.

  Let's see how AWS IoT Core handles authentication and authorization for those IoT devices.

For authentication, there are two types required:

- **Server authentication**: IoT devices need to make sure they are sending the data to the right, legitimate IoT Core endpoint, hence, AWS IoT Core should authenticate itself to those IoT devices. Server authentication is usually done by using a digital certificate. AWS IoT Core (that is, the server) has two X.509 certificates: one certificate signed by Verisign, which is legacy now, and another certificate signed by Amazon Trust Services (the preferred option by AWS). You need to configure the IoT devices that will connect to AWS IoT Core to have Verisign or an Amazon Trust Services certificate stored in those IoT device trust store locations. In other words, when IoT devices connect to or establish a TLS connection with AWS IoT Core, they will be presented with an AWS IoT Core certificate (either the one signed by Verisign or the one signed by Amazon Trust Services – it depends on which endpoint you use). Now, those IoT devices need to validate those certificates to make sure they are talking to AWS IoT Core and nothing else. That validation can be done by having the public key of the entity that signed those certificates (Verisign or AWS Trust Services) stored in those IoT devices.

- **Client authentication**: For this authentication type, IoT devices need to authenticate themselves to AWS IoT Core. Under this category, we have the following methods supported by AWS IoT Core:

  - If you use an AWS IoT Core HTTPS endpoint or you use MQTT over HTTPS/WebSocket, then AWS IoT Core offers authentication using AWS SigV4. SigV4 is a process to add authentication information to AWS API requests sent over HTTP. Usually, this SigV4 is done behind the scenes. If you use AWS SDKs, you just provide your AWS access key and the SDKs will handle the SigV4 process. If you are not using SDKs, then you have to do the signing process yourself when you call the AWS APIs.

  - If you use an AWS IoT Core HTTPS endpoint or MQTT over HTTPS/WebSocket and you want to use another method for authentication instead of SigV4, then AWS IoT Core gives you the flexibility in building that custom authentication workflow by using solution building blocks such as AWS Lambda (that is, where you will write your authentication custom logic or code) and a custom authorizer component where you define the authorization function, that is, the Lambda function; the `tokenKeyName`, which is the name of the parameter the device will send in the custom authorization HTTP request header; and finally, the `tokensigningpublic` key, which you use to sign the token you generated. You can check the AWS online documentation for full details on that custom authorization method.

- If you use an AWS IoT Core MQTT (over TCP/TLS) endpoint, then AWS IoT Core provides mutual authentication using an X.509 client certificate. This is the most commonly used authentication method in production-grade IoT solutions. It is the recommended authentication method due to the fact that the key that the IoT device uses in authentication (that is, the private key) never leaves the IoT device, meaning it is super secure.

Let's deep dive into this mutual authentication option as it is important.

We assume you have the required basic knowledge and understanding of X.509 certificates that use asymmetric cryptography or the public-key cryptography method; however, here is a quick refresher on digital certificates:

- In communication between two entities, if you have only one encryption key used for encrypting and decrypting some data, then this is called a symmetric key and both entities, that is, the sender and receiver, have to have the same encryption key.

- Asymmetric encryption is another encryption concept where there are always two keys paired together, one used for encryption and the other one used for decryption. Both keys only work with each other.

- Asymmetric keys solve the challenge of sharing the encryption key as you share only one of the pairs of keys you have (that is, the public key) and you keep the other key secured (that is, the private key) and not shared with anyone. You can publicly share your public key – hence the name "public." There's no issue in that as you hold the private key that is paired with that public key. You (as the holder of the private key) are the only one who can decrypt any messages encrypted by your public key.

- Digital certificates solve another problem or challenge. Yes, public-private key encryption is very secure, but you still can have a man -in -the -middle type of security attack where someone sits in the middle between the true sender and the true receiver and pretends they are the legitimate true sender. They then share with the legitimate receiver their (that is, the attacker's) public key. The true receiver will encrypt the information using that false public key, and then the middle entity (that is, the attacker) will decrypt the message since it is encrypted using their public key. From the other end, the middle entity (that is, the attacker) acts as the legitimate receiver and uses the right public key of the legitimate sender to encrypt the message and send it as if they are the true receiver, and so on.

A digital certificate can solve that man -in -the -middle attack as the public key of the legitimate entity is now shared within a certificate that is signed by a trusted **Certificate Authority (CA)** confirming that the public key shared in the certificate is related to entity X. You now know you are talking to the right entity (entity X) in the communication and not to a false or pretend one.

Public keys are shared in what is called a certificate, but how is this certificate signed, or, to be more accurate, how is it digitally signed? The answer is certificates are also digitally signed using the public-private key concept. The signer of the certificate holds a private key securely and its public key is widely known and trusted by operating systems, browsers, and so on.

The entity that signs the certificate is called the CA.

There are public CAs such as DigiCert, GoDaddy, and Let's Encrypt. Those public CAs have public keys (or their self-signed certificate as they are root CAs) and usually come installed by default in operating systems, browsers, and so on.

You send your certificate to those public CAs containing your public key and they sign that certificate with their private key. In the browser, for example, the browser will receive your server certificate from your web server during the TLS handshake. The browser or the operating system has the public key of the CA (or the certificate that holds the CA's public key) that signs that certificate, meaning it will confirm that your certificate is valid.

You can build your own private CA. A certificate signed by a public CA is not free, so using a publicly signed CA certificate in massive IoT devices might not be the best option (that is, expensive).

For each device, you will need to have a unique certificate installed on that device. This is recommended from a security perspective, so if the device certificate is compromised, you can simply revoke that certificate from a single device or disconnect the device. This is better than impacting all your devices if you use one single certificate for all your devices.

- There are root CAs (these are entities that self-signed their certificate) and there are intermediate CAs, which are other entities that also act as CAs, but their certificate is signed by the root CA or by another intermediate CA that is signed at the end by a root CA.

Mutual authentication means clients, which are IoT devices in our case, authenticate the server (IoT Core), that is, they make sure they are talking to the right IoT Core. We covered that part in the server authentication section. Now, the server, in return, needs to authenticate those clients, that is, IoT devices.

We mentioned the fact that the IoT devices will use X.509 certificates to authenticate themselves to IoT Core. Now, let's understand how the X.509 certificates that will be used by the IoT devices will be created and authenticated by AWS IoT Core.

AWS IoT Core provides the following options for IoT device certificates:

- **One-click device certificate generation**: In that option, AWS will create a public and private key and will create a certificate holding that public key and then sign that certificate by an AWS IoT CA.

  You just download the signed device certificate and its private key.

  > **Note**
  > Creating a public key, private key, certificate, and so on is easy. A tool such as OpenSSL is usually used in creating those files.

  - In this option, that is, one-click device certificate generation, AWS IoT Core does everything on your behalf. Note that the device certificate, in the end, is signed or trusted by the AWS IoT CA, hence, when you send that certificate to AWS IoT Core, AWS IoT Core will be able to validate and authenticate it as it is signed and trusted by the AWS IoT CA.

  - There are some steps to be done in AWS IoT Core, including registering and activating the generated device certificate in AWS IoT Core. You simply click on a button to generate the device certificate, and then you download the generated files (that is, the device certificate, public key, and private key). Then, you activate that device certificate in AWS IoT Core. Now, AWS IoT Core is ready to authenticate your device certificate.

  - This option is usually used in development or POC projects, not in production-grade IoT projects. See the next option, which is better than this option but is still not the best.

- **Provide your own certificate signing request (CSR)**: In that option, you create the public, private key, and CSR or the non-signed certificate that holds your public key. Then, you send that CSR to AWS IoT Core to be signed by the AWS IoT CA. AWS will sign your CSR and then you download the signed device certificate that is signed by the AWS IoT CA.

  The benefit of that option is you hold the private key and never share it with AWS IoT Core. In the one-click option, AWS generates everything for you, including the private key. Yes, AWS might not store your private key and it is generated on the fly, but it's better to generate the private key yourself anyway and never share it with anyone.

Note that in that option, you still use the AWS IoT CA as a trust for your device certificates, so you still need to download the certificate after it is signed by the AWS IoT CA. You don't have control over the certificate management process, that is, if you need to revoke or invalidate the device certificate, and so on. This option might not be used in large-scale and production-grade IoT solutions, but it can be used in small IoT projects.

• **Bring your own certificate (BYOC)**: In that option, you do everything yourself and you have full control. So you start by creating a public key, private key, and CSR, and then you sign that CSR with your own trusted CA. With that, you will have a device certificate signed by your private trusted CA, but AWS IoT uses another trusted CA, that is, the AWS IoT CA, hence, it will not be able to validate/authenticate your device certificate unless AWS IoT Core trusts your private trusted CA.

You upload your trusted private CA certificate to AWS IoT and then AWS IoT will use your private trusted CA to trust device certificates that are signed by your private trusted CA.

This option is the most widely used option in large-scale and production-grade IoT solutions as you can handle device certificate generation, signing, and so on during a device manufacturing process, for example.

If you have your own trusted CA, then you can create intermediate CAs. For example, if you have different product lines, for each product line, you can create an intermediate CA. If you want to discontinue one product line, for example, you can simply revoke or make the intermediate CA of that product line untrusted. In short, this option has much more flexibility in production-grade IoT solutions.

There are two processes, called just-in-time registration and just-in-time provisioning. We will cover these in the next chapter.

Now that we are done with the first step, which is authentication, the next step is authorization. We need to know how AWS IoT Core handles the authorization part, or what authenticated IoT devices are allowed and not allowed to do in AWS IoT Core.

The AWS Cloud in general uses **authorization policies** to manage authorization. Policies are JSON documents that contain a set of permissions. A policy contains one or more policy statements. Each statement contains the following parameters (or keys):

• **Effect**: The value of this key can be Allow or Deny, which means the policy allows or denies access.

• **Action**: The value of this key holds the list of actions that the policy allows or denies.

In AWS IoT, actions can include `iot:Connect` and `iot:Subscribe`.

- **Resource**: The value of this key specifies the resources to which the actions are applied.

In AWS IoT, a resource can be a thing object, which represents the actual physical IoT device in AWS IoT Core.

An IoT Core authorization policy is attached to a thing. In fact, you attach the IoT Core policy to the certificate that is registered in AWS IoT Core (if you use an X.509 client certificate for authentication), attach it to the AWS IAM policy (if you use MQTT over WebSocket or HTTPS using AWS SigV4 for authentication), or attach it to a custom authorizer policy in the case of custom authentication.

To summarize, whatever authentication method you will use, the next step by AWS IoT Core will be checking the AWS IoT Core policy or AWS IAM policy to evaluate what IoT devices can and cannot do:

- **Device Gateway**: This component is the entry point for IoT devices connecting to AWS IoT Core. It provides a bi-directional communication model where it securely receives traffic from IoT devices and helps in sending traffic to IoT devices. It is a fully managed and scalable gateway that supports over a billion devices and messages. AWS Device Gateway currently supports MQTT, WebSocket, and HTTP 1.1 protocols.

- **Message broker**: This component is the AWS implementation of the MQTT message broker. IoT devices, applications, services, and systems can publish and subscribe to the different topics hosted in the message broker.

The message broker supports MQTT, MQTT over WebSocket, and HTTPS. The way you connect to the message broker depends on the protocol you are using. Different protocols require different authentication mechanisms, as explained earlier.

- **Rules engine**: This is one of the most important components in AWS IoT Core (and any other IoT platform as well) as it is the glue between the physical world and the cloud world. IoT data generated by sensors is sent across to the IoT Core message broker, and then it is the responsibility of the rule engine to move that data out from the message broker to the right destination in the cloud for further processing.

There are many destinations configured in the rule engine – see *Figure 5.4*. Note that the screenshot doesn't show all the destination services that the rule engine can be integrated with.

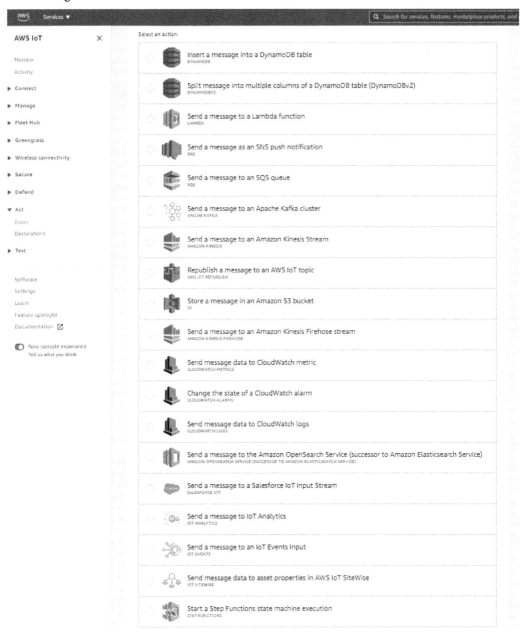

Figure 5.5 – AWS IoT Core – rule engine

The rule engine uses SQL-like statements to filter and route MQTT messages. The following example shows a rule configured to read all MQTT messages published to the cityA/parkingA topic, for example, and insert those messages into an S3 bucket called cityAparkingA:

```
{
awsIotSqlVersion": "2016-03-23",
sql": "SELECT * FROM 'cityA/parkingA'",
ruleDisabled": false,
actions": [
s3": {
"roleArn": "arn:aws:iam::11111111:role/aws_iot_s3",
"bucketName": "cityAparkingA",
"key": "myS3Key"
}
}
]
}
```

- **Device Shadow**: This is one of the interesting IoT services in AWS IoT Core. A device shadow is a JSON document that virtually represents the state of the actual IoT device.

  The device shadow document holds the current state and the desired state of the IoT device.

  Device shadow services were introduced to address the challenge of intermittent internet connectivity issues.

  With a device shadow, if the IoT device went offline, the apps would still be able to get the status of those IoT devices and still be able to send commands to those IoT devices. Typically, apps are integrated with an IoT device shadow, not directly with IoT devices.

  Apps and IoT devices interact with the cloud-based device shadow service, so if an app wants to retrieve the state of an IoT device, it will interact with the device shadow service to retrieve the current state of the IoT device, which might not be the actual/real state but the last reported state of the device when it was online.

Apps can change the IoT device state to another state, that is, the desired state. Apps interact with the device shadow service to make that change even if the actual IoT device is offline. The device shadow service will buffer the commands and will send them to the IoT device when the device is back online again.

Apps or IoT devices can interact with a device shadow through MQTT topics or HTTP REST APIs. There are reserved MQTT topics for the shadow service.

The device shadow document is associated with a thing object that represents the actual IoT device in AWS IoT Core.

The device shadow document contains some metadata information and a state property that describes the device's state. Under the state property, there are nested properties:

- `desired`: This property defines the desired state of the IoT device, for example, `fan_switch":"off` or `fan_switch":"on`. Only apps or services update this property.

- `reported`: This property holds the device's reported state. Only the IoT device updates this property.

- `delta`: This property is updated by AWS IoT to report the differences between desired and reported states.

The device shadow concept exists on most IoT platforms on the market. In Azure IoT, it is called `digital twins`.

This was a quick overview of AWS IoT Core and its main components. In the AWS documentation, you will find lots of tutorials covering each component in detail. We focused more on the concepts, as they are almost the same on any other IoT platform that you might select for your IoT backend.

The preceding AWS IoT Core service is mainly for internet-connected devices. In other words, it is for devices that have internet connectivity. AWS recently introduced a new service called AWS IoT Core for LoRaWAN, which enables customers to connect wireless devices that use low-power, **long-range wide area network (LoRaWAN)** technology to the AWS IoT platform.

AWS provides a managed **LoRaWAN Network Server (LNS)**. Check *Chapter 2, The "I" in IoT – IoT Connectivity*, for more information about LoRaWAN. AWS IoT Core for LoRaWAN includes support for an open source LoRaWAN gateway-LNS communication protocol called Basics Station. Once LoRaWAN gateways and devices are connected, device data is automatically routed to the AWS IoT Core rule engine, accelerating IoT application development.

In the next section, let's briefly go through some of the other AWS IoT services that we will cover in this chapter.

Other AWS IoT services

In our earlier roadmap of AWS IoT platform services, we mentioned some AWS IoT services that will be explained in different chapters of this book and some other services we will explain briefly in this chapter besides the AWS IoT Core service.

Here are quick descriptions of those other AWS IoT services that might be of interest to you:

- **AWS IoT 1-Click**: This service is usually used for quick IoT prototyping, PoC projects, or the quick launch of a simple IoT product.

  The idea of the AWS IoT 1-Click service is to have a ready-made, simple, ready-provisioned, and easily programmable IoT device that is already connected to AWS IoT.

  IoT 1-Click devices are usually simple, small, and have buttons and LEDs; you can program them in the cloud (that is, Lambda functions) using whatever logic you want to be executed when the user presses or clicks the buttons of that device. Such devices come already pre-provisioned with a certificate to make it easy to connect to AWS IoT Core. They also have a ready-made internet connectivity option, which could be Wi-Fi or cellular connectivity.

  Using this setup, you can program anything with just a press of a button on this device and you don't need to worry about device firmware, connectivity, IoT Core, MQTT, and so on; you just need to write the Lambda function in Java, Python, or C#, for example, which will be triggered on the press of a button. Some devices offer *long-press*, *short-press*, *single-click*, *double-click*, and so on, so you can program different logic accordingly.

  Most of the projects we've seen using that service have mainly been prototyping projects; however, some companies, such as mobile network operators, partner with AWS and other device manufacturers to produce such IoT-ready devices using their cellular connectivity SIMs and sell the devices to IoT developers to enable them to build their IoT prototyping projects. The cost of such IoT-ready devices is cheap and you can buy them from Amazon.com or from other vendors' marketplaces.

- **AWS IoT Things Graph**: This service mainly helps non-developer users to rapidly build IoT applications by combining devices and web services and defining the interactions between them with little or no code.

It offers a visual drag-and-drop user interface for connecting devices and web services to build IoT applications that can run on the cloud or the edge (AWS Greengrass).

It has powerful workflow management to automatically coordinate and track each step of the flow you build.

It represents devices and services using pre-built reusable components, called models, that abstract low-level details such as communication protocols, message syntax, and the interface.

It comes with pre-built models for common devices such as lights, motion sensors, and **Programmable Logic Controllers** (**PLCs**), or you can create your own custom model.

It also has built-in tools to monitor application performance, set alarms, and view logs.

Summary

In this chapter, you learned about IoT platforms and their capabilities, the solution building blocks that IoT platforms provide, and how to select the best IoT platform for your IoT solutions.

You also learned about MQTT, the most common and well-known IoT application communication protocol.

To get a practical sense of IoT platforms, we took a deep dive into one of those IoT platforms – the AWS IoT platform. In this chapter, you learned about some of the services offered by the AWS IoT platform, including AWS IoT Core, AWS IoT 1-Click, and AWS IoT Things Graph.

In the next chapter, we will continue our IoT journey by covering the IoT device management domain, and we will also cover some of the AWS IoT platform services in that domain, such as AWS IoT Device Management and IoT Device Defender.

6
Understanding IoT Device Management

IoT device management is such an essential IoT service that you'd find it hard to find a large-scale and production-grade IoT solution that doesn't use it.

In large-scale IoT solutions, you need to have full control and management over deployed IoT devices, which are typically in the thousands or even millions. If you don't have an efficient way to manage and control that fleet of devices, then many issues can affect your IoT business – operational, support, bad customer experience, and security issues, to mention just a few.

An IoT device management capability offers a broad spectrum of functionalities that will support any IoT solution, such as **device provisioning and authentication, device configuration and control, device monitoring, device diagnostics and troubleshooting**, and **firmware and software updates and maintenance**.

In *Chapter 1, Introduction to the IoT – The Big Picture*, we explained the different IoT solution design patterns that are used in most large-scale IoT solutions. An IoT device management solution or capability is an implementation of the IoT command design pattern, which we explained briefly in *Chapter 1, Introduction to the IoT – The Big Picture*.

Some device management solutions implement or offer capabilities for the telemetry and command IoT patterns, but in this book, and this chapter in particular, we will focus only on the command IoT design pattern as part of the IoT device management solution, since we covered the telemetry pattern in the previous chapter when we talked about the IoT core component.

In this chapter, we will follow the same methodology we followed in the previous chapters; we will start talking about generic concepts relating to the topic, the different vendors of the device management solution, and how you can select the best IoT device management solution for your IoT solution. Then, we will continue our journey through the AWS IoT platform; this time, we will cover the **AWS IoT Device Management** services.

In this chapter, we will cover the following topics:

- An IoT device management overview
- AWS IoT Device Management
- AWS IoT Device Defender

An IoT device management overview

Managing and controlling remote-connected devices is not a new concept; our mobile phone devices, for example, are remotely managed. With mobile phones, whether they are smart or non-smart, mobile operators can push some configurations to the end user's mobile phone device through an SMS bearer, for example, to configure the mobile phone's internet access point or any other features in the device. Mobile operators or administrators can also manage the whole mobile device of the end user remotely.

So, the concept is not new, but IoT brings some challenges to that concept, such as the largescale (that is, a massive number of connected IoT devices that need to be managed), lowpower (that is, the devices are in sleep or deep-sleep mode most of the time, so devices are not on or connected most of the time), and lowcost of IoT devices (that is, the devices are constrained and limited in resources).

Conceptually, and from a solution point of view, the way a device management solution typically works is that you have a client or agent installed on the IoT device (that is, the endpoint IoT device or gateway/Edge IoT device) and the control systems or applications (aka the device management system) sitting remotely in the cloud.

The device control system will communicate with the device agent in a bidirectional way where the device will respond to instructions sent by the device control system. Given that simple architecture paradigm, you can implement your device management solution.

There are many communication protocols that you can use for managing devices, such as **Lightweight Machine to Machine (LwM2M)**, **MQTT**, **Modbus**, **Controller Area Network Bus** (**CANBus**), and **REST**.

Before we deep dive into the technical aspects of device management solutions, let's first understand the features provided and enabled by an IoT device management solution:

- **Device onboarding and provisioning**: In the previous chapter, we talked about creating an IoT thing in Cloud IoT Core. The IoT thing is a virtual representation of an actual IoT device. We talked about creating a unique identity for the IoT thing/ IoT device using the X.509 certificate or secret keys, and so on. Those identities are used to authenticate the IoT devices when they connect to IoT Core. There are many steps required to onboard or provision the IoT devices first into the IoT platform before they start normal IoT operations, such as sending and receiving IoT data and messages with Cloud IoT Core.

  Device management solutions streamline, automate, and simplify the IoT devices' onboarding and provisioning into an IoT platform. The typical IoT device management features provided in this category are **single- or bulk-device** provisioning, device bootstrapping, **Just-in-Time Provisioning** (**JITP**), and registration (aka on-demand device provisioning).

- **Software updates and maintenance**: This is a very important feature of any proper IoT device management solution, as it enables you to patch and update the software running on the IoT devices or update the device firmware to a newer version. This makes your IoT device future-proof and flexible to any new changes required.

  Consider a scenario where IoT devices are already deployed or sold with very old and vulnerable firmware. The customers or end users who bought those devices will be using out-of-date firmware, which you will have to update remotely to a newer, more secure, and stable version.

  Updating firmware or applications of IoT devices remotely is also known as **Over-the-Air** (**OTA**) updates.

- **Device configuration management**: This feature enables device operators or administrators to control and configure a device remotely – for example, devices might initially configure to connect to the IoT cloud on a specific endpoint. You might need to change such configurations on the device. An IoT device management solution enables you to do that and read all the device configuration that is already in place.

- **Device diagnostics and troubleshooting**: This feature enables you to troubleshoot device connectivity issues or device issues in general. A device management solution runs some diagnostic tools and checks to report back to you the state of each check (step) so that you can easily identify the source of the device problem. This could be an issue with the battery, connectivity, software/operating system, and so on.

These IoT device management features bring the following business benefits to your IoT solution:

- **IoT operational excellence and efficiency**: With an IoT device management solution, you will be fully aware of your IoT device fleet status and what is happening on those devices. You will be alerted immediately if something happens on the devices, and you will reduce the unexpected device downtime.

- **Reduce the total cost of ownership**: With IoT device management solutions, engineers don't need to physically visit the site where devices are deployed to update or manage them. Also, there will be a reduction in the operational efforts and cost for updating and managing those IoT devices.

- **IoT device optimization**: With IoT device management features such as OTA updates, remote device monitoring, and device configuration management, you can keep your IoT devices up to date in terms of device performance, security, and rapid deployment of your new IoT business features.

We mentioned earlier that there are many protocols used for IoT device management. The most used protocols in that domain in IoT device management markets are LwM2M and MQTT. We covered MQTT in the previous chapter, so let's discuss the LwM2M protocol briefly in the next section, as it is an important device management protocol and standard that an IoT solution architect or designer should be aware of.

LwM2M

LwM2M is a device management protocol defined and standardized by **Open Mobile Alliance (OMA)** SpecWorks. OMA SpecWorks is a standardization body and the original creator of the **OMA-DM (Open Mobile Alliance – Device Management)** protocol, which allowed mobile operators to manage mobile phones, PDAs, and tablets more efficiently.

With lots of demand for IoT and machine-to-machine communication, OMA SpecWorks comes with the LwM2M protocol, which is considered a next-generation and industry-leading IoT device management protocol.

LwM2M is similar to MQTT to some extent, as both are standard and open protocols for device and system communications, lightweight, and fit for low-power and resource-constrained devices that operate over low-bandwidth networks.

LwM2M is different from MQTT in the following ways:

- **Architecture**: MQTT follows **publish/subscribe (pub/sub)**-based communication, while LwM2M follows client-server-based communication. In LwM2M, IoT devices act as LwM2M clients that interact with the LwM2M server to register, bootstrap, configure, and report the devices.

  LwM2M relies on or is built on top of **Constrained Application Protocol (CoAP)**, which is a specialized IoT application protocol designed for communication in constrained and limited devices. CoAP provides a request/response interaction model between application endpoints and follows the REST paradigm (that is, resources represented as URIs).

  MQTT runs on top of TCP/IP, while LwM2M runs on top of UDP and many other means of transport.

- **Purpose**: **Message Queuing Telemetry Transport (MQTT)**, as the name suggests, is a telemetry protocol, so it was designed for telemetry purposes, not device management purposes. However, MQTT is used by some vendors in implementing device management solution features, but it is up to the vendor management how they will implement these device management features. To give an example, you can use the MQTT building blocks to implement a firmware update feature or any other device management feature, but they will be custom-built from scratch, and the implementation of those features might be different between device management vendors.

  Alternatively, LwM2M was organically and specifically defined and designed as a device management protocol. It gives a list of predefined and standard interfaces, objects or data models, and resources that cover device management features such as firmware updates, device monitoring, device provisioning, and bootstrapping.

> **Important Note**
> There are many standardization working groups that work with LwM2M specifications. They define protocol specification, interfaces, and what is known as a device object or device data model.

The device can support different objects; the specifications define the object format that the device management vendors must adhere to in order to be classified as an LwM2M-compliant device management solution.

Each object has a global assigned ID, and it contains one or more resources. A resource can be readable, writable, or executable. The resource can be individual or in a list, mandatory or optional, and have a specific data type.

To give an example, one of the **Internet Protocol for Smart Objects Alliance (IPSO)** Smart Objects is a temperature object. That temperature object has many resources (each resource has a unique ID); here are some of the temperature object resources to give you an idea:

- **Sensor value**: This resource represents the current measured temperature sensor value.

- **Minimum measured value**: This resource represents the minimum value measured by the temperature sensor since power was switched on.

- **Maximum measured value**: This resource represents the maximum value measured by the temperature sensor since power was switched on.

- **Sensor units**: This resource represents the measurement unit used, such as Cel for Celsius.

The sensor value resource, for example, has a unique ID, is mandatory, has a float data type, and is a readable resource.

So, the preceding temperature object and its resource are what come defined by standards (that is, the IPSO), so if the device management vendors want to add additional resources or different temperature objects, they can work with the IPSO Working Group to define that.

As mentioned earlier, LwM2M is built upon CoAP, which uses the REST architecture paradigm, so IPSO Smart Objects or a data model will be represented in the URI of LwM2M interfaces/APIs (as discussed shortly) with the following format – object ID/object instance ID/resource ID.

For the temperature object we used as an example, the object ID will be the ID of the temperature sensor, the object instance ID will be a specific instance of the temperature sensor object, and finally, the resource ID will be the resource you are looking for.

There are many predefined and ready objects coming from different working groups working in LwM2M specification standards (OMNA, IPSO Smart Objects, and GSMA objects).

Here are some of the objects (with an object ID range of 0-499) defined by the OMA and DMSE working groups:

- **LwM2M security (ID: 0)**: This object holds the keying material of the LwM2M client to access the LwM2M server.

- **LwM2M server (ID: 1)**: This object holds data related to the LwM2M server.

- **LwM2M access control (ID: 2)**: This object holds authorization information to check the access rights for different operations.

- **Device (ID: 3)**: This object holds information related to a device that can be queried by the LwM2M server, including device reboot and factory reset functions.

- **Connectivity monitoring (ID:4)**: This object holds information related to the network connectivity of the device.

- **Firmware update (ID:5)**: This object is for the firmware update capability.

- **Location (ID:6)**: This object is for device location information.

- **Connectivity Statistics (ID:7)**: This object holds information such as transmit and receive counters.

IPSO Smart Objects offer lots of predefined objects, such as temperature (as covered earlier), humidity, power measurement, actuation, light control, an accelerometer, a magnetometer, and a barometer.

For a full list of all objects from all the LwM2M standards groups, check out the OMA **LwM2M** Object and Resource Registry online.

The LWM2M specification also defines the following interfaces:

- **Bootstrap interface**: This interface is exposed to configure devices with the required security and credentials that they need to connect to the LwM2M server. The flow is simple – the device connects to the LwM2M server, asking for its credentials; the server, after doing some checks on the initial credentials shared by the device, replies with device credentials that the device should use to connect to the LwM2M server.

- **Registration interface**: This interface enables devices to register/de-register to or from the LwM2M server. The flows in that interface are also straightforward, as the device connects to the LwM2M server and asks for registering, de-registering, and so on.

- **Device management and service enablement**: This interface enables the LwM2M server to retrieve and query objects that the device supports, retrieve the device status, invoke a device and actuator actions, and so on. The flow of the interface starts from the LwM2M server calling the devices.

- **Information reporting**: This interface enables the LwM2M server to be notified when the values of device-supported objects change. The flow of the interface starts with the LwM2M server first configuring the device for the notification threshold and objects that are under observation. When the threshold occurs, the device will call back LwM2M to notify it.

> **Important Note**
>
> With such well-defined and standardized data models (objects) and interfaces, device management vendors can implement and offer an LwM2M-compliant device management solution. This is one of the greatest benefits of using LwM2M in device management solutions instead of MQTT, as LwM2M offers more interoperability and plug-and-play architecture, while device management solutions with MQTT will always be dependent on custom vendor implementation.

Why is interoperability so important in device management solutions? In large-scale and production-grade IoT solutions, IoT devices might come from different device vendors and manufacturers, so if you mandate that all devices should support well-defined standards such as LwM2M, then you can simply and quickly onboard and provision those devices into your LwM2M device management solution and start exploring and using the different device management features, such as firmware updates and device monitoring.

The question now is which one to use for the IoT device management solution – LwM2M or MQTT?

Without any doubt, LwM2M has the upper hand, as it is specifically designed to serve the device management capability, but sometimes, you might have to accept a device management solution that doesn't support LwM2M, which could happen when you have already bought or have most of your IoT backend cloud services, such as IoT Core and IoT Analytics, running with a specific vendor that also offers a device management solution based on MQTT and not LwM2M. At this point, you will need to make a trade-off architectural decision between the benefits you get from an LwM2M-compliant (such as interoperability) and an MQTT-compliant (such as good integration with other IoT services, since the vendor is the same, cheaper, and quicker to use) device management solution.

To give a practical example regarding the preceding point, the AWS IoT platform offers IoT platform services including AWS IoT Core and AWS IoT Analytics. It offers AWS IoT Device Management solutions that are MQTT-based, not LwM2M. If you have already done the exercise for selecting an IoT platform, as explained in *Chapter 5*, *Exploring IoT Platforms*, and you have decided to go with the AWS IoT platform, then you will see that selecting an AWS IoT Device Management solution will be the practical and pragmatic option, unless you have business requirements that mandate using a device management solution that is compliant with and supports LwM2M standards.

The previous point assumes you have selected the AWS IoT platform and you don't have an existing LwM2M device management solution that you have been asked to use as your IoT device management solution. Although the pragmatic and practical option in this case is to go with AWS IoT Device Management, what if you have an IoT device management solution that is LwM2M-compliant and you have decided to go with the AWS IoT platform for other IoT services, such as IoT Core? In that case, the answer is to use and build a bridge (from MQTT to LwM2M and vice versa) between your existing LwM2M server and AWS IoT Core services, so that your IoT devices use LwM2M for provisioning, bootstrapping, and so on – in other words, they use LwM2M for control and command features and AWS IoT Core for IoT data ingestion and telemetry. This solution will require custom implementation, as there's nothing resembling this bridge concept provided out of the box.

In the next section, let's see a list of device management solution vendors and how to select a device management solution for your IoT project.

Device management vendors and the selection criteria

There are so many device management vendors on the market, such as **Cumulocity**, **Telit**, **Sierra Wireless**, **PTC**, **Microsoft**, **Amazon**, **Nokia**, **Bosch**, **ThingsBoard**, and **Particle**.

The following are the criteria that you should consider and evaluate when selecting an IoT device management solution for your IoT solution:

- **Interoperability and openness**: What kind of protocols and standards does the candidate IoT device management solution support and offer? Does it support a standard device management protocol such as LwM2M or does it use MQTT or other proprietary protocols in building its device management solution? Is it an open source or commercial solution?

- **Device management feature set**: The more the candidate device management solution offers different device management features that cover all device life cycle management (from device manufacturing to the device's **End of Life** (**EOL**), the more business and technical value you will get from that device management solution. Typical features provided by an IoT device management solution include, but are not limited to, software updates and firmware management, monitoring, alerting, device provisioning, diagnostics, logging, troubleshooting, and remote configuration.

- **Integration**: An IoT device management solution is one component out of many in any large-scale IoT solution technology landscape; hence, it is important to choose an IoT device management solution that offers rich and complete integration with other IoT solution building blocks and components used in a project, so you need to check whether the candidate device management solution offers well -defined APIs and SDKs or not, whether it provides rich developer documentation and development toolkits or not, and so on.

- **Architecture**: It is important to check the architecture of the candidate IoT device management solution and whether it is multitenant, scalable, extendable, secure, and easy to use and integrate with or not.

- **Connectivity management**: It is also important to check how the candidate IoT device management solution handles and manages the connectivity of the device, as without proper connectivity management, you will lose control of the device and accordingly lose all device management solution features.

In the next section, we'll discuss one example of an IoT device management solution, which is AWS IoT Device Management.

Our Recommendation

You should have – as much as possible – all those selection criteria covered by the selected IoT device management vendor, but we know in reality it is usually not so easy, especially in large-scale IoT projects where you not only have a device management solution to worry about but also other solution building blocks that you should consider as well in your decision. Therefore, you usually – as an IoT solution architect – end up doing some architecture and business trade-off to decide upon a best-fit solution in the end.

AWS IoT Device Management

AWS IoT Device Management is one of the AWS IoT platform services that AWS offers to enable IoT customers to onboard, organize, monitor, and remotely manage their IoT -connected devices at scale.

Broadly, AWS IoT Device Management provides the following services or features:

- **Device provisioning and registration**: This feature covers different IoT device provisioning and registration flows into AWS IoT Core, such as single or bulk provisioning and auto-device registration and provisioning (such as Just-in-Time Registration JITR and JITP). You'll find more on this feature in the following sections.

- **Fleet indexing and searching services**: This feature enables customers to gain more visibility and insights into the health and status of their device fleet. The indexing and searching service depends and counts on two main sources – the AWS IoT device registry, where the metadata of the IoT devices or IoT things are stored (as explained in the previous chapter), and the AWS IoT Device Shadow service, where the real-time status of those IoT devices is stored.

  By combining these two types of data sources (that is, device registry and Device Shadow), the search queries will be powerful and real-time driven, so instead of just having a static query (that is, device registry data), such as `find all devices with manufacture X`, you can have a real-time query (that is, with Device Shadow's help), such as find all devices with manufacture X that had firmware version 1.0 and are currently connected.

  This indexing feature also helps in building dynamic device grouping or organization so that you can have a dynamic group based on a query such as group all devices with firmware version 2.0 that are located in London. This means that every time a new device is added to the fleet with the same condition, it will be automatically placed in that group.

  Device grouping helps with another feature that we will discuss next – jobs.

- **Jobs**: This is the AWS implementation for the device software and firmware updates OTA. AWS, behind the scenes, uses MQTT to implement that feature.

  You can target a job to a group of devices or a specific single device, which gives you much more flexibility in the deployment of new software or firmware versions to your IoT devices as, with that feature, you can make updates in batches to reduce the blast radius of a failure that could happen because of the new updates.

AWS also allows you to digitally sign the jobs before they are sent to your devices to protect them from being compromised. We will look further into this feature in the following sections.

- **Fleet Hub**: This service helps to build standalone web applications to monitor the health of the IoT device fleets. You can make these web applications available to the end users in your organization, even if they don't have AWS accounts. The Fleet Hub service has many capabilities, such as monitoring device fleets in near real time, setting alerts, and running IoT device jobs.

- **Fine-grained device logging and monitoring**: This feature helps with device monitoring and troubleshooting by collecting device logs and streaming them into the Amazon CloudWatch service for search in those device logs or built custom metrics and alarms in CloudWatch. The AWS IoT platform has many metrics of the different services it provides, such as AWS IoT Core and AWS IoT Device Management.

  With this feature, you can configure the logging level on a per -device basis or a group of devices, and you can also increase the diagnostic level across devices.

Before we deep dive into the AWS IoT Device Management features and services, you need to know that you can use the AWS IoT console to do some basic service setup and configuration, but with SDKs/APIs, you can do a much more advanced service setup and implement the device software logic (the device agent) on an IoT device. AWS provides the following options:

- **AWS IoT Device SDKs**: This includes open source libraries, developer guides, and sample apps that enable you to build innovative IoT solutions on both the device side and the server or application side. The SDKs come in C++, Python, JavaScript, Java, and Embedded C.

- **AWS Mobile SDKs**: This provides mobile app developers with platform support for the APIs of the AWS IoT Core services, IoT device communication using MQTT, and the APIs of other AWS services.

- **AWS IoT Device Client**: The AWS IoT Device Client is free, open source, modular software written in C++ that you can compile and install on your embedded Linux-based IoT devices to access the AWS IoT CoreAWS IoT Device ManagementAWS IoT Core, AWS IoT Device Management, and AWS IoT Device Defender AWS IoT Device Defender features by default. It serves as a reference implementation for your IoT devices to work with AWS IoT services, with operational best practices baked in – using the AWS IoT Device Client is the easiest way to create a **Proof of Concept** (**PoC**) for your IoT project. What's more, since it is open source, you can modify it to fit your business needs, or optimize it when you wish to scale up from a PoC to production (`https://github.com/awslabs/aws-iot-device-client`).

> **Important Note**
>
> In the next few sections, when we talk about the device agent in the IoT device, we mean broadly the AWS IoT Device Client that is installed on the IoT device. You can use other SDKs provided by AWS, such as IoT Device SDKs, to build the device agent yourself, but that might take time and will not be as efficient as the AWS IoT Device Client.

Now, let's explain the device provisioning and jobs features in more detail in the next few sections, but bear in mind that although we will explain the features from an AWS IoT Device Management solution perspective, the concepts are the same even if you go with other vendors or implement a device management solution yourself. Therefore, it is still important to understand the AWS IoT Device Management features.

Device provisioning and registration

The device provisioning feature allows you to register your IoT devices individually or in bulk. In the previous chapter, we explained the main building blocks required for registering or provisioning an IoT device into the AWS IoT platform; the building blocks are as follows:

- **An IoT thing**, which is a virtual representation of the actual IoT physical device. We explained thing types and thing groups in the previous chapter.

- **An IoT device certificate**, which is used for authentication with AWS IoT Core. We explained this in the IoT in the previous chapter as well.

- **An IoT policy**, which is attached to the IoT device certificate. The IoT policy outlines what the IoT device is and is not allowed to do in AWS IoT Core.

These are the three main components that you need to create to register or provision an IoT device in AWS IoT before the device can start communicating with AWS IoT Core.

All of what we will subsequently explain concerns how efficiently and smoothly these three components can be created in the AWS IoT platform.

Before we dive into it, note that registration is just one step of the provisioning flow. The device certificate is registered in AWS IoT, but this doesn't mean the IoT device is now fully provisioned and can start communicating with IoT Core; you will still need to activate the certificate and attach the policy to it to mark the IoT device as fully provisioned in the AWS IoT platform.

We can classify the IoT device provisioning process in AWS into two categories:

- **In-advance device provisioning**

- **On-demand or automatic provisioning**

With in-advance device provisioning, you can provision a single device or bulk devices. If you want to provision one device at a time, you can use the single-device provisioning flow. However, in production-grade and large-scale IoT solutions, you will usually provision the IoT devices in bulk, not one by one.

AWS provides provisioning APIs and what is known as a device provisioning template to enable both types of device provisioning.

The device provisioning template is a JSON document that outlines the resources you need in the provisioning process or flow – basically, the three components mentioned earlier: the IoT thing, the IoT certificate, and the IoT policy.

The concept of a device provisioning template works for both single- and bulk-device provisioning, but it is more valuable in bulk-device provisioning where you define your things (attributes, the thing type, and so on) and provide a parameter file in the JSON format that holds all the values of the list of devices you need to provision. The parameter file contains a list of name-value pairs for the parameters used in the provisioning template, and the file can be stored in an S3 bucket.

If you check the AWS IoT RegisterThing API, for example, you will see it takes the provisioning template and parameters as input.

The in-advance approach works perfectly if you already have a well-known IoT device with its attributes and characteristics known and categorized beforehand.

On-demand provisioning offers capabilities to provision IoT devices when they are connecting to AWS IoT for the first time or when you want to provision the IoT devices into AWS IoT but you don't have enough information about them yet to register them as bulk or even single provisioning.

For on-demand device provisioning, AWS offers many options:

- **JITP**
- **JITR**
- Fleet provisioning by a trusted user
- Fleet provisioning by claim

We explained in the previous chapter that the X.509 certificate is the most common device identity option used in large-scale and production-grade IoT solutions. We also explained how to sign such a device certificate, and we mentioned that the best option is to use your own **Certificate Authority** (**CA**) certificate to sign the IoT device certificates.

Now, if you manage to store the device certificate that is signed by your CA certificate and its private key in a secure location in the IoT device, either during the device manufacturing process or afterward, then you can use **JITP** or **JITR** to dynamically provision that IoT device into AWS IoT Core. But how?

Think about that scenario – an IoT device has an X.509 certificate signed by your CA certificate that is stored and installed on the device, which is not yet provisioned into the AWS IoT platform. In other words, you didn't create an IoT thing, IoT certificate, and IoT policy for it in AWS IoT Core; you just registered your CA certificate (the certificate that is signed by the IoT device certificate) with AWS IoT and enabled what is called auto-registration on that CA certificate. Now, the IoT device powers on and tries to establish the first connection with AWS IoT Core, which will challenge – during the TLS handshake – the IoT device to present its certificate to authenticate it. The IoT device presents the device certificate, but AWS IoT has no idea about that device certificate, as it is not provisioned yet in AWS IoT, but it is signed by your CA certificate, which has already been registered and validated in AWS IoT. Hence, AWS IoT will realize the auto-registration feature is enabled for that CA certificate, so it will realize that the device tried to be dynamically provisioned. AWS IoT will then trigger one of the on-demand device provisioning flows we mentioned earlier (JITP and JITR).

So, the first TLS handshake will clearly fail, but it is used to trigger the on-demand device provisioning flow. Then, the IoT device should have logic to reconnect again to AWS IoT Core, but this time the device will be already provisioned and the TLS handshake will succeed, and the device will start communicating with AWS IoT Core.

Let's see the difference between JITR and JITP in the next sections.

JITR

We'll start with JITR; as this feature was launched before the JITP feature, you will see that JITR requires some work to be done by you to complete the device provisioning flow, while JITP does everything and eliminates the additional steps you did in JITR.

The steps for JITR are as follows:

1. On the AWS IoT side, create, register, and activate the CA certificate that will be used or has already been used to sign the IoT device (that is, the IoT endpoint or IoT Edge device) certificate.

2. On the AWS IoT side, enable the auto-registration of certificates.

   Step 1 and *step 2* are required in JITR and JITP; they are important steps.

3. On the IoT device side, you need to create an X.509 certificate and sign it with the registered CA certificate. You also need to program the device to reconnect after a while when the TLS handshake with AWS IoT Core fails for the first time.

   On the AWS IoT side, create a rule in the AWS IoT rules engine with the `Topic` filter:

   ```
   $aws/events/certificates/registered/<caCertificateID>
   ```

   Replace `<caCertificateId>` with the ID of the registered CA certificate and set the action for that IoT rule to be a Lambda function; let's call it, for now, an auto-provisioning Lambda function.

   When the IoT device attempts to connect to AWS IoT Core with an X.509 certificate that is not known to AWS IoT but was signed by a CA that was registered with AWS IoT, the device certificate will be auto-registered by AWS IoT in a new `PENDING_ACTIVATION` state. You need to change the status of the registered certificate from `PENDING_ACTIVATION` to `ACTIVE` so that it can be successfully authenticated; otherwise, if the device continues to connect while its certificate is in a `PENDING_ACTIVATION` state, the connection will fail.

   AWS IoT, when setting the certificate state in `PENDING_ACTIVATION`, also triggers a device certificate registration event by publishing a message to the following reserved MQTT topic:

   ```
   $aws/events/certificates/registered/<caCertificateID>
   ```

   Here, `caCertificateID` is the ID of your registered CA certificate that is registered in AWS IoT and used for signing the IoT device certificate.

The message published on this topic has the following structure:

```
{
    "certificateId": "<certificateID>",
    "caCertificateId": "<caCertificateId>",
    "timestamp": "<timestamp>",
    "certificateStatus": "PENDING_ACTIVATION",
    "awsAccountId": "<awsAccountId>",
    "certificateRegistrationTimestamp":
  "<certificateRegistrationTimestamp>"
}
```

Now, you can simply build a new rule in the AWS IoT rules engine to subscribe to that topic:

```
$aws/events/certificates/registered/<caCertificateID>
```

Then, you can do whatever logic you want by calling the Lambda function, as explained to do the device provisioning.

4. Create a Lambda function (auto-provisioning) in whatever language you want that is supported by AWS Lambda (Node.js, C#, and so on). This Lambda function will implement the IoT device provisioning flow, which includes creating an IoT thing and an IoT certificate, creating and attaching an IoT policy to the IoT certificate, and activating the IoT certificate. There is a sample of the auto-provisioning Lambda function code in Node.js, available in the `aws-sample` repository on GitHub: `https://github.com/aws-samples/aws-iot-examples/blob/master/justInTimeRegistration/deviceActivation.js`.

With this device provisioning flow, any new device can be provisioned automatically in the AWS IoT platform and start communicating with AWS IoT Core, given that the new device has a device certificate that is signed by a CA certificate that is registered and has auto-registration enabled in the AWS IoT platform.

From a security point of view, if the IoT device certificate is compromised, you can revoke that IoT device certificate in AWS IoT, but if the CA certificate itself is compromised, then you can deactivate/deregister the CA certificate in AWS IoT. The connection will fail when IoT devices – signed by that compromised CA certificate – try to communicate with AWS IoT Core because the CA certificate is deactivated (has entered an inactive state), so AWS IoT Core will not allow that connection.

> **Important Note**
>
> The greatest benefit of the JITR device provisioning flow is that you have full control of how the IoT device will be auto-provisioned into the AWS IoT platform. You can subscribe to the events published by AWS IoT, which are MQTT messages that are published to well-known preserved AWS IoT topics. By using the AWS IoT rules engine, you can deliver those messages to other AWS or non-AWS services to do the custom provisioning logic you might have. In the example we mentioned, we used AWS Lambda as an action for the IoT rule, but nothing stops you from choosing any other action or service for that rule.

JITP

JITP was the new feature introduced by AWS IoT to eliminate the need to create a Lambda function and an IoT Core rule to automatically provision IoT devices into AWS IoT Core.

The JITP flow is like the JITR flow; basically, steps 1 to 3 in the JITR flow are also used in the **JITP** flow with only a very small change – while you are registering the CA certificate in AWS IoT, in the same command (`register-ca-certificate`), you will associate a provisioning template, so you simply pass the additional `--registration-config` file://provisioning-template.json argument to that registration command. Note that the argument used to enable auto-registration in that command is `--allow-auto-registration`.

The main difference between JITR and JITP are steps 4 and 5, as they are not needed in the JITP flow. You don't need to create a Lambda function (auto-provisioning) and create an IoT rule; now, the AWS JITP workflow will do that on your behalf. You just need to define the device provisioning template and associate it with the registered CA certificate, as explained earlier.

The provisioning template attached to the registered CA certificate can have or reference the following parameters:

```
AWS::IoT::Certificate::Country
AWS::IoT::Certificate::Organization
AWS::IoT::Certificate::OrganizationalUnit
AWS::IoT::Certificate::DistinguishedNameQualifier
AWS::IoT::Certificate::StateName
AWS::IoT::Certificate::CommonName
AWS::IoT::Certificate::SerialNumber
```

The value of these parameters can be extracted – at runtime – from the subject field of the IoT device certificate that is trying to get provisioned into AWS IoT.

The JITP workflow will substitute the references of these parameters with the values extracted from the certificate and will provision the resources specified in the provisioning template accordingly.

The JITP workflow will create the following resources, as specified in the attached provisioning template:

- An IoT thing
- An IoT policy
- An IoT certificate

Then, the JITP workflow will do the following:

- Attach the created IoT policy to the certificate.
- Attach the certificate to the created IoT thing.
- Update the certificate status to ACTIVE.

JITR and JITP workflows work perfectly when IoT devices have unique credentials in the form of a client certificate. These can be stored in the devices during or after the manufacturing process, but how can we provision IoT devices that do not have such a certificate installed on them at all, either during or after manufacturing? In that case, there could be no certificate installed on those IoT devices because of some technical limitation, cost, or any other reason.

AWS offers a service called fleet provisioning and offers two ways to provision devices with some sort of unique credentials after those IoT devices are delivered to the end user. The two ways are fleet provisioning by a trusted user and fleet provisioning by claim.

Let's explore those fleet provisioning types in the next couple of sections.

Device provisioning by a trusted user

The idea in this provisioning workflow is simple – the end user (device operator or administrator) will be initially authenticated and trusted instead of the IoT device itself, so the device installer will use a mobile app that will authenticate with AWS, and if authenticated successfully, the installer will get a temporary X.509 certificate and a private key that is valid for 5 minutes. These temporary credentials must be delivered and stored in the IoT device, which must have the capability to accept the temporary credentials over a secure connection, such as Wi-Fi, Bluetooth Low Energy, or even USB.

The IoT devices must also implement logic to publish and subscribe to the fleet provisioning of well-known and reserved MQTT topics.

With the temporary credentials, the device will start to communicate with AWS IoT by publishing a message to `$aws/certificates/create/json`. The fleet provisioning service will create a new certificate and a private key signed by the AWS CA and publish a message with the certificate, private key, and certificate ownership token to the `$aws/certificates/created/json/accepted` topic.

Now that the device has got permanent credentials, it will start the on-demand device provisioning process. It will publish the required parameters and certificate ownership token it got earlier to `$aws/provisioning-template/{template-name}/provision/json`.

From the other end, the fleet provisioning service will listen to that topic and start creating the required resources based on the provisioning template. Once the resources are created and the device is provisioned in AWS IoT, the fleet provisioning service will publish a message to `$aws/provisioning-templates//{template-name}/provision/json/accepted`.

The device should have logic to disconnect and start a new session with AWS IoT Core and connect normally.

That is a great feature, but what if devices do not have the capability to accept credentials over secure transport or can't be customized to accept and store credentials after they have been manufactured?

To answer this question, AWS provides a fleet provisioning by claim service for that purpose, which we will discuss in more detail in the next section.

Device provisioning by claim

In this workflow, you initially create a certificate and its associated private keys and embed them in the firmware that will be used with your IoT devices. Let's call that certificate the default certificate for now.

On the AWS IoT side, you must register and activate the default certificate, which will only be used for the first connection with AWS IoT Core. As part of this provisioning flow, the IoT device will get a permanent certificate that is signed by the AWS CA `root` certificate, and it will use the permanent certificate for the subsequent connections to AWS IoT Core as normal.

On the AWS IoT side, you also need to create a restricted IoT policy that is attached to the default certificate to restrict the use of such a claim certificate (that is, a default certificate) – for example, you can define specific fleet provisioning topics such as `/$aws/certificates/create/*` and `/$aws/provisioning-templates/<templateName>/provision/*` that the device can publish and subscribe to.

So, the temporary certificate (aka the claim certificate) that comes by default with device firmware will only be used to get a permanent and unique certificate for the IoT device.

As with on-demand device provisioning, you will create a device provisioning template that will be used to provision the IoT devices into AWS IoT Core.

The provisioning by claim flow will be as follows:

1. The device will connect for the first time to AWS IoT Core using the claim certificate (the default certificate) that is already registered and activated in AWS IoT Core, so the connection will be accepted. However, remember that at that moment, the device can't publish or subscribe to any topic other than the topics defined in the IoT policy that is attached to the default certificate. Earlier, we mentioned that the IoT policy will restrict the device to publishing and subscribing to fleet provisioning topics only.

2. The device has logic to publish a message to `/$aws/certificates/create/json`. The fleet provisioning services listen to that topic and create a new certificate and private key signed by the AWS CA and publish the newly created certificate, private key, and certificate ownership token to the `/$aws/certificates/create/json/accepted` topic. The device will listen or subscribe to the `/$aws/certificates/create/json/accepted` topic, so it will retrieve a new certificate – that is, a permanent client certificate and a private key – and store them securely in the device.

3. Now, the device will start on-demand device provisioning by publishing a message with all the required parameters, certificate ownership, and so on to `/$aws/provisioning-templates/<templateName>/provision/json`. The fleet provisioning service listens to that topic and triggers the Preprovisioning hook Lambda function to perform some additional verification logic if needed. If everything is fine, the Lambda function should return `{"allowProvisioning": True}` to the fleet provisioning service, which will accordingly create the required resources for the IoT device, such as the IoT thing and the IoT policy, and activate the IoT device certificate based on the provisioning template. With that, the device provisioning is complete.

4. By now, the device has already been provisioned and can start communicating with IoT Core using a unique permanent certificate.

In the preceding IoT device provisioning flows, we talked about IoT devices that are provisioned into AWS IoT Core in one single AWS Region within one single AWS account, but what if you need more flexibility in where those IoT devices should be provisioned?

For example, you might sell your IoT devices globally; hence, you need the devices that are sold in Europe to be provisioned only into the AWS IoT Core endpoint in AWS European Regions, such as Frankfurt, London, or Dublin, and the IoT devices that are sold in South America to be provisioned only into the AWS IoT Core endpoint in AWS South American Regions, such as São Paulo. Alternatively, you might have different AWS accounts used for developing, testing, and production, so you might need a way to move the IoT-provisioned devices from one account to another in the same AWS Region or different Regions for some reason (for example, if you don't need to register the CA certificate in AWS IoT). If so, AWS provides what is known as **Multi-Account Registration (MAR)**.

To use MAR, you must not register the CA certificate that is signed by the IoT device certificate in AWS IoT; you just need to register the IoT device certificate in all AWS accounts or Regions that you are targeting for your business. Finally, the IoT devices must have logic to use the correct `host_name` value in the SNI extension (the AWS IoT Core endpoint URI is different in each Region) when it initiates the TLS connection with AWS IoT.

With that, we have covered the different device provisioning and registration flows that are available and supported by AWS IoT Device Management. In the next section, we will talk about another interesting feature in AWS IoT Device Management, which is AWS IoT Jobs.

AWS IoT Jobs

Jobs is an AWS IoT Device Management feature that enables remote device operations such as reboots, factory resets, certificate rotation, remote troubleshooting, firmware, and software updates.

To create a job, you must create a job document that defines the remote operations that should be done on a device. The job document is a UTF-8 -encoded JSON document and can be stored in an S3 bucket or passed in line with the command that is used to create the job.

Usually, job documents contain some instructions and URLs of the software binaries or configuration files that the device will download as part of the job execution and completion – for example, if you want to upgrade the device firmware, you upload the new version of the firmware software binary to an S3 bucket and create a job document containing the instructions of how to update the firmware and a URL of where the binary of the new firmware is available. The agent on the device will execute the required command, download the new version of the firmware from an S3 bucket, and so on.

On the device side, there must be an agent installed to communicate with the AWS IoT Device Management cloud Jobs service. The communication is done based on the MQTT protocol, and the agent software should be able to interpret the job document and execute what is instructed in that document.

The steps are straightforward, as follows:

1. Create a job document containing the instructions of what you want to apply to the devices. It is better to store the job document in an S3 bucket; you should also store other files you expect the device agent to download in an S3 bucket.

2. On the **AWS IoT side**, there are some predefined jobs you can create, such as creating a FreeRTOS OTA update job, creating a Greengrass V1 core update job, or creating a custom job, which means you implement the job yourself on the device side.

3. On the **AWS IoT side**, create a job and define the targets of it. The targets are the IoT devices that the job will be applied to. The targets can be things, thing groups, or both. The AWS IoT Jobs service will send a message to each target to inform it that a job is available for the device to be executed.

4. When creating a custom job on the AWS IoT side, you will select the job file, where you will simply be asked about the S3 URL of that file.

The AWS IoT Device Management Jobs service and the device agent installed on the device communicate through specific, reserved, and well-known MQTT topics, such as `$aws/events/job/`, and `$aws/events/jobExecution/`:

- **On the IoT device side**, the agent will execute instructions in the job document and communicate to the AWS IoT Device Management Jobs service the status of the job execution.

- **On the AWS IoT side**, you can track the job execution either by checking the job status in the **AWS IoT console** | **manage** | **Jobs** menu, or you can use the MQTT test client that comes with the AWS IoT console (or any other client) to subscribe to relevant job MQTT topics and track the job status.

There are some job configurations you should also be aware of when creating a job:

- **Job run type**: There are two run types of jobs that you can select from. The first is called snapshot, which means the job is marked as completed after it deploys to all of the job targets. The other job run type is called continuous, which means the job will continue to deploy to the new devices that are added to the thing groups in the job target list.

- **Rollouts**: You can specify how quickly the targets are notified of the pending jobs, which enables you to create a staged rollout to better control your fleet updates to not update all the devices at the same time, and so on.

- **Aborts**: You can create a set of conditions to abort rollouts when the criteria you define are met – for example, abort the job if 20% of the targets, of which there are 100 in total, have their job status as TIMED_OUT.

- **Timeouts**: You can define how you will be notified whenever a job execution gets stuck in IN_PROGRESS for an unexpectedly long period of time.

Here are some notes on the AWS Device Management Jobs feature:

- AWS Jobs uses the MQTT protocol (not LwM2M) to implement different device management features, as previously explained.

- As explained earlier, and concerning the previous point – that is, the usage of MQTT and not LwM2M – with the AWS Jobs service, you implement the job document yourself so that you can do whatever you want with the device operations and software updates.

In *Chapter 1, Introduction to the IoT – The Big Picture*, we explained the device outbound connectivity-only IoT design pattern and that it is recommended for IoT devices – from a security point of view – to not open their inbound interfaces or ports. In other words, it is better to not access an IoT device directly through its inbound port, so the IoT device will usually do outbound connectivity only.

We saw in this chapter that the IoT design pattern has been applied perfectly with services such as AWS IoT Device Management Jobs, where a device will connect to well-known and reserved topics to retrieve the jobs that need to be run on it. We saw the same pattern applied when we talked about the AWS IoT Device Management provisioning services. But, in some cases, you might need to interactively access the IoT device synchronously to do, for example, some troubleshooting on the device, configuration updates, and other operational tasks. So, the question now is, can we access the device synchronously without compromising the security practices in place? In other words, is there a way to access the device remotely without opening its inbound ports and changing the existing firewall policy that protects the device?

AWS answers that question by offering a service called **AWS IoT secure tunneling**. Let's see how this service works in the next section.

AWS IoT secure tunneling

The AWS IoT secure tunneling service enables AWS IoT customers to establish bi-directional communication with remote IoT devices over a secure connection that is managed by AWS IoT. AWS secure tunneling does not require updates to an IoT device's existing inbound firewall rules.

To explain how this service works, let's understand the different concepts and components used in the AWS IoT secure tunneling service:

- **IoT device administrator**: This is the person who wants to access the IoT device remotely.

- **Administrator workstation (source device)**: This is the device or the workstation that is used by the IoT device administrator. This can be a laptop or desktop PC (the source device) that the administrator uses to initiate a session to the destination device (IoT device). The source device (that is, the administrator workstation) must run the LocalProxy software. The LocalProxy software will be discussed further shortly.

- **IoT device (destination device)**: This is the IoT device that the IoT device administrator wants to access. This must run the LocalProxy software and an IoT device agent that can integrate with AWS IoT Core and subscribe to tunnel notification topics over MQTT.

- **Tunnel**: This is a logical pathway through AWS IoT to enable bi-directional communication between source and destination devices.

- **Client access token**: To establish a secure tunnel between the source and destination devices, the AWS IoT secure tunneling service generates a pair of tokens, one to be used by the source device when it connects to the tunnel and the other for use by the destination device when it connects to the tunnel from its end.

- **LocalProxy**: This is a software proxy that runs on both the source and destination devices to relay the data stream between the secure tunneling service and the applications of the devices on both sides.

> **Important Note**
>
> The tunneling concept is implemented behind the scenes using the WebSocket protocol. This is why you need to make sure that the IoT device can initiate traffic (outbound) toward port 443 (HTTPS) so that the LocalProxy initiates the WebSocket session with AWS IoT Device Management to establish a secure tunnel connection from both sides – that is, the source and destination devices – and it will relay the data stream on both sides (there is a LocalProxy run on both sides for this reason).

- **Destination application**: This is the application that runs on the destination devices (the IoT devices) that need to be accessed by the source device (that is, the workstation administrator) – for example, the IoT device should have an SSH daemon running in the IoT device if the source device wants to access the daemon in the IoT device to establish an SSH session with the IoT device. SSH is just an example of an application that can be accessed remotely; you can access other services in the IoT device, but you would need to configure them.

Now, let's see how the preceding components interact and work together to establish secure tunneling between the source device and the destination device:

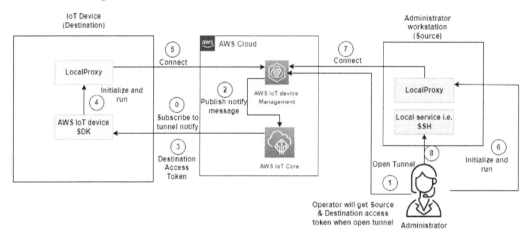

Figure 6.1 – AWS IoT secure tunneling

The flow of establishing a secure tunnel between the administrator workstation device and the IoT device is follows. The following step numbers correspond with the circled numbers in the diagram:

- Step 0: In this step, the IoT device is configured to subscribe to the $aws/things/thing-name/tunnels/notify tunnel notification topic to always check whether there are any open tunnel requests or not.

The main objective of this step is to obtain the **destination access token** that is required in the IoT device to open the tunnel from its side by initiating and starting the LocalProxy software in the IoT device to start the connection with the AWS IoT secure tunnel.

This step might not be required if the device administrator manages to send or transfer the destination access token to the IoT device and start the LocalProxy software either manually or automatically in the IoT device to establish the secure tunnel connection. This is why we count this step as *step 0*.

- Step 1: The flow starts when the device administrator creates or issues a command to open a secure tunnel in AWS IoT Device Management through the AWS IoT console or APIs.

When the device administrator issues a command to open the tunnel, they will get a pair of access tokens – the aforementioned destination access token, which is meant to be sent to the IoT device (that is, the destination device), and the other, called a **source access token**, which is meant to be used by the administrator workstation device (that is, the source device). The traffic will be initiated from the source device (that is, the administrator workstation device).

- Step 2: Once the open tunnel command is completed in the previous step, AWS IoT Device Management will publish a tunnel notification message to the tunnel notification topic (see *step 0*).

- Step 3: The IoT device agent that is subscribed to the tunnel notification topic will receive the message that AWS IoT Device Management has published, as per the previous step.

It is recommended to have the AWS IoT Device Client installed on the IoT device as it supports the AWS IoT secure tunneling feature and has an embedded LocalProxy, so you don't need to install the LocalProxy separately; just install the AWS IoT Device Client on the IoT device.

- Step 4: The IoT device client will retrieve the destination access token from the tunnel notification message, as in the previous step, and then will initial and start the LocalProxy software. As explained before, that step might not be needed if the administrator manually pushes the destination access token to the IoT device and starts the LocalProxy software.

- Step 5: The LocalProxy (in destination mode) will establish a connection from its destination side with the AWS IoT secure tunneling service. After this step, the destination device (that is, the IoT device) is now connected, but the source device (that is, the administrator workstation device) is not yet connected.

- Step 6: The administrator already has the **source access token**, as per *step 1*, and they can now see that the destination device is connected to the secure tunnel. It is now time for the source device (that is, the administrator workstation device) to connect from its end to the secure tunnel, so the administrator will initiate and run the LocalProxy (*in source mode*) in the administrator workstation device to establish the connection with the AWS IoT Device Management secure tunneling service from the source side (the administrator workstation device).

- Step 7: Once the connection is established from the source device toward the AWS IoT secure tunnel, the connection is completed, and the administrator can access the services running on the destination device (that is, the IoT device).

- Step 8: The administrator workstation device can now access services such as SSH that run on the IoT devices by issuing an SSH command to the LocalProxy locally. The LocalProxy will relay that command to the IoT device through the tunnel (through the WebSocket). The LocalProxy in the IoT device will receive the command and pass it to the local running service in the IoT device (that is, the SSH daemon).

The administrator can configure any service in the IoT device to access, but SSH is usually used for accessing the IoT device remotely and starting device troubleshooting and investigation.

With this, we have completed the AWS IoT Device Management solution to some extent, so let's talk about some other services that are related to AWS IoT Device Management. In the next section, we will talk about the AWS IoT Device Defender service.

IoT Device Defender

As the name suggests, this service is a fully managed device security service that helps in securing the IoT devices that are onboarded and provisioned in the AWS IoT platform.

The AWS IoT Device Defender service continuously audits the IoT devices against IoT configurations that you define to make sure the devices are kept secure and are not deviating from the security best practices.

So, broadly, we can say that the AWS IoT Device Defender service offers the following features:

- **Audit device configuration for security vulnerabilities**: AWS IoT, continuously or on demand, audits the IoT device-related resources, such as a device's X.509 certificates, IoT policies, connection settings, and account settings, against the AWS IoT security best practices. To give an example, the audit checks whether the principle of least privilege is used or not, whether there is a shared certificate used with IoT devices or not, whether conflicting MQTT client IDs are used or not, whether the CA certificate is expiring, inactive, revoked, or not, whether the device certificate is expiring, revoked, inactive, or not, and many other checks.

- **Continuously monitor the device behavior to identify anomalies**: The AWS IoT Device Defender service monitors high-value security metrics from the cloud (AWS IoT Core) and IoT devices and compares those metrics against the expected device behavior – for example, you usually expect the device to have 15 connection attempt messages every day with AWS IoT Core. If that number increases or decreases, that is a sign that something wrong happened, and then you can act accordingly.

  If you already know the device's behavior, then you just need to create what is known as a rule-based anomaly detection security profile, such as the aforementioned example about the expected number of IoT device connection messages per day. However, what if you don't know the device's expected behavior? AWS IoT Device Defender answers that by offering a **Machine Learning (ML)** Detect security profile. That profile utilizes standard ML models, such as the anomaly detection model, to proactively figure out any anomalies that occurred in the device.

So, in short, if you know the device's behavior, use a rule-based anomaly detection security profile; if you don't, then use the ML Detect security profile.

AWS IoT Device Defender ML Detect uses metrics to detect the anomalous behavior of devices. It compares the reported value of the metric with the expected value you provide (in the case of rule-based anomaly detection). With ML detection, it uses anomaly confidence thresholds (high, medium, and low) for the metric-based behavior.

So, metrics are an important element of the AWS IoT Device Defender service, and there are two types of metrics that are currently supported:

- **Cloud-side metrics**: These kinds of metrics are generated on the cloud side (the AWS IoT cloud), such as the number of authorization failures, the number or size of messages a device sends or receives through AWS IoT, connection attempts, disconnects, and source IP.

- **Device-side metrics**: These kinds of metrics are generated by the device and sent to the cloud. To use device-side metrics, you can use the AWS IoT SDK on your AWS IoT-connected devices or device gateways to collect the metrics and send them to AWS IoT. Bytes in, bytes out, the established TCP connection count, packets in, packets out, and so on are examples of device-side metrics that you can use while creating an audit security profile in AWS IoT Device Defender.

The AWS IoT Device Defender service publishes security alarms to the AWS IoT console, Amazon CloudWatch, and Amazon SNS when the audit fails or when anomaly detection occurs.

Alarms will only be sent to SNS if it is configured to be on. If alarms are configured to be suppressed, the AWS IoT Device Defender service will continue flagging anomalous behavior as violation alarms. However, suppressed alarms won't be forwarded for SNS notification. They can only be accessed through the AWS IoT console or API.

Alerting or alarming is a great feature, but to complete the audit cycle, the AWS IoT Device Defender service offers mitigation actions to minimize the impact of the security issue that was detected. AWS IoT Device Defender provides predefined mitigation actions, such as replacing a default policy version, disabling a device certificate, enabling IoT logging, and quarantining a device. You can also define some custom mitigation actions, such as restarting a device, pushing a security patch, or any other mitigation action you might have. Custom mitigation actions can be done by subscribing to the AWS SNS topic that AWS IoT Device Defender publishes audit check results to, so you can act accordingly upon receiving those messages from the SNS topic.

To summarize, the AWS IoT Device Defender service defines and applies the typical audit, check, alarm, and act or mitigate IoT device defense life cycle. The steps to implement this cycle in the AWS IoT Device Defender service are simple: you start by defining a security profile (using the AWS IoT console or APIs) that defines anomalous behavior for a group of devices (a thing group) or all devices in your AWS account, which enables you to specify what actions to take when an anomaly is detected. The anomalous behavior can be based on rule-based or ML detection. AWS IoT Device Defender uses security-related data metrics collected on the cloud and device sides along with the behaviors defined in the security profile to detect anomalies in the behavior of IoT devices.

Summary

In this chapter, you have learned about IoT device management platforms and solutions, what capabilities and solution building blocks they provide, and how to select the best IoT device management platform or solution for your IoT solutions.

You have also learned about the different device management protocols that are used and supported in IoT device management solutions. You have learned how the LwM2M protocol is the next-generation and industry-leading IoT device management protocol. You also learned how the MQTT protocol is used to build a device management solution besides being used mainly for telemetry purposes.

To get some practical sense of IoT device management solutions, we explored one of them, AWS IoT Device Management. We learned about the different AWS IoT Device Management features, such as device provisioning and registration, fleet indexing and searching services, Fleet Hub, AWS IoT Jobs, and AWS IoT secure tunneling.

You also learned about the IoT Device Defender security service and how that fully managed service helps in protecting and securing IoT devices at scale.

In the next chapter, we will continue our IoT journey, and this time, we will cover the IoT analytics domain and some of the AWS IoT platform services covering that domain, such as AWS IoT Analytics, AWS IoT Events, and AWS IoT SiteWise.

7
In the End, It Is All about Data, Isn't it?

Open any modern **big data** or **advanced data analytics** book or reference and without any doubt, you will find the phrase **Internet of Things (IoT)** mentioned frequently. Do you know why? Because IoT is considered one of the greatest data sources for the big data domain.

As stated in the earlier chapters, no one buys technologies just for the sake of playing with them. Technologies are invented and keep evolving to solve business problems. The main purpose of IoT technologies, in the end, is to use valuable real-world data to gain business value and insights; this is what IoT technologies provide to business stakeholders.

Collecting data is one benefit of IoT technologies, and another is managing and controlling IoT devices remotely. For the latter benefit, some might argue that managing devices remotely has already been available for decades now – mobile phone devices are a great example – so accordingly, we can say that the ultimate benefit of IoT technologies is collecting data from the real world.

Data is very important in modern business. You may have already heard one or more of the following buzzwords: data is the new oil, data is the fuel for business digital transformation, data is a product, data-driven organization. All of these stress the importance of data.

Successful business enterprises follow what is known as a customer-obsession or customer-first principle or strategy. They always try to collect and analyze as much data as they can about their customers to offer the best experience to them. The business also gets other benefits such as reduced operational cost (predictive maintenance, near real-time monitoring and notifications, and so on), increased revenue (selling and recommending new products and services, and forecasting customer needs), and protection of the business position/brand in a competitive market.

There are many data points that can be collected about customers from different data sources such as **Customer Relationship Management (CRM) systems**, **billing systems**, **customer support systems** (that is, ticketing systems), **customer services systems** (that is, call centers), **customer online interactions**, and many other data sources about customers. By adding IoT data to that list, enterprises or businesses will have a complete and true picture of their customers, or what is known as a 360-degree customer view.

Think about a scenario where a customer has a problem with a product they bought from company X. Earlier, the customer made a call to company X's customer service support team to complain about some issues with the product they bought. However, with the help of IoT data, the customer does not need to call the customer service support team to complain, as the company will already be fully aware of the customer's problems and will proactively call them to inform them about the expected issues that they could face with the product they have. Also, the company might arrange with that customer a replacement or a fixing service for the expected issues – interesting, isn't it? Enterprises use massive amounts of data generated from IoT devices and other sources to leverage advanced analytics and **Machine Learning** (**ML**) techniques such as predictive maintenance and anomaly detection, as we will explain in this chapter.

> **Note**
> It is important to understand that IoT data is a subset of big data, so designing an IoT data analytics solution is almost the same as designing a big data and advanced data analytics solution.

Big data or **advanced data analytics**, **ML**, and **Artificial Intelligence** (**AI**) are big and complex topics that require entire books to be fully covered and explained, so clearly, this chapter, and this book in general, cannot be used by any means as a reference for covering those type of technologies, as they are out of scope.

Our main goal in this chapter is to simply help IoT solution architects or designers think about and devise an architecture and design for an IoT data analytics solution for their large-scale and production-grade IoT projects. It is rare to find a large-scale IoT project without a data analytics component or solution represented in an E2E IoT solution architecture.

We will cover some concepts and solution building blocks of generic data analytics solutions, and then we will cover AWS IoT Analytics and other AWS services that are commonly used in the data analytics domain.

In this chapter, we will cover the following topics:

- An IoT data analytics overview
- Data analytics solution building blocks
- The AWS IoT Analytics service
- Other AWS services commonly used with AWS IoT Analytics

An IoT data analytics overview

Traditional data analytics or **Business Intelligence** (**BI**) solutions have been available for decades now. They follow the standard and well-known data analytics or data mining process that starts with data extraction from source systems. This is followed by the data transformation process and then loading data into purpose-fit data analytics stores, such as SQL-based data warehouses or a relational database. This process is usually called **Extract, Transform, and Load** (**ETL**).

Traditional data analytics is different from modern or **advanced data analytics**; in traditional data analytics, a business analyst or business owner starts by already having data from different data sources. Then, they will ask the question, OK I have all the data – what kind of information will I get out of such raw data? Then, they ask the question, Now that I have the information from the raw data, what kind of business insights will I get out of such valuable information? Finally, with those business insights gained, their next question will be, Based on these valuable business insights, what business actions or recommendations should I implement to achieve the desired business goals or benefits?

Alternatively, in advanced or modern data analytics, it is the opposite. The business owner starts with the question, As a business owner, what kind of actions or recommendations should we implement to increase our revenue, reduce operational cost, and maintain our position in the market? Then, they work backward by saying, Well, to achieve such business goals, we need to first have some business insights or metrics. Then, the next question will be, Well, to get such business insights, what kind of information should we have in place first? Finally, as you might have guessed, the last question will be, Well, to get such information, what kind of data should we collect?

So, your first and most important job is to define the data that the business needs to collect to get the required insights and business benefits it is looking for.

During the ETL process or pipeline, source or row data will get cleaned, transformed, enriched, and finally, stored in a purpose-fit analytics store, which is ready for the analytics engine to access and run analytics queries to create reports and dashboards to gain business insights from the raw data.

Traditional data analytics was based mainly on **batch processing**, which means at the end of the business day (or half-day, every couple of hours, every week, or whatever timescale that is not real time or near real time), a background process was run to collect or extract raw data from the different data sources and transform it into a proper format, which was used to efficiently run analytics queries such as data warehouses.

IoT data is classified as streaming data, meaning that it is append-only data; in other words, it is not like a transactional database that supports updates and deletions of the data after it gets inserted into a database, so if the IoT device sends an event or data stream that the temperature at 5:30 P.M. was 40 degrees, that fact will never, ever change! Therefore, always remember that IoT telematics data is append-only data; in other words, IoT data is a perfect example of time-series data where you can easily represent the time in the x axis (or the y axis) and the IoT data generated at a timestamp in the other dimension (the y axis or the x axis).

IoT time-series data can be regular (that is, data coming continuously in second or minute intervals) or it can be irregular (that is, data coming as events that are triggered when something has happened).

> **Important Note**
> IoT data is huge and constantly growing in volume, variety, and velocity, which is why IoT data analytics is somehow classified under the umbrella of big data analytics.

The business benefits of IoT data analytics

There are many business benefits to applying IoT or data analytics in general, such as the following:

- **Predictive maintenance**: This is the most common business benefit of applying IoT data analytics. IoT devices usually have many sensors attached to them, and you continuously get streams of data about the device behavior and the surrounding environment (such as weather); hence, with some ML models or even traditional analytics methods, you can predict any issues that could occur in those IoT devices – for example, if the temperature of the device keeps increasing, that might be a signal or an indicator of some fault happening in the device, meaning it might be about to break.

- **Recommendations and forecasting customer needs**: Based on customer behavior, you can recommend to a customer another service or product that is cheaper than what they currently use – for example, in the telecoms sector, customers might purchase a mobile price plan with lots of data allowance and less SMS allowance. If you notice that the customer uses SMS more than data, then you could recommend that they move to another product or service that is cheaper than the one they are currently on and that fits perfectly with their needs.

- **A new revenue stream**: Data as a product is the new trend, and nowadays, there are many companies just selling data in what is known as data marketplaces.

- **Real-time monitoring and alerting**: This is a very important business benefit that is applied in many business sectors. In the healthcare sector, for example, the medical devices attached to a patient's body may have sensors that continuously report patient biometrics such as temperature and heart rate, so a healthcare professional can get alerted immediately and in real time if some trend occurs or a threshold is reached, allowing them to intervene quickly if needed. Also, this type of patient data with some advanced ML models can predict diseases such as cancer and diabetes. Real-time monitoring and alerting are also used heavily in manufacturing, industrial, and other sectors.

Let's understand the different types of data analytics in the next section.

Types of data analytics

As an IoT solution architect designing a data analytics platform for your IoT project, the first thing you should do is ask yourself or the business stakeholders what the problem is you are trying to solve. Is IoT data analytics the solution for that business problem? What business value will be gained by applying IoT data analytics? What kind of data should be collected and why? What kind of data analytics is required (descriptive, diagnostic, predictive, or prescriptive)?

Let's briefly explain those types of data analytics:

- **Descriptive analytics**: This type of data analytics focuses mainly on what has happened/is happening. It deals with what has happened in the past based on the data collected. The data analytic outcome usually shows – in dashboards/ visualizations – the different business **Key Performance Indicators** (**KPIs**), metrics, statistics, and reports.

  To give an example, in a smart parking system, you can ask questions such as what the average parking lot occupancy per hour, day, week, or month is in a specific parking zone, a specific city, or all cities.

- **Diagnostic analytics**: This type of data analytics focuses on why something happens. In the smart parking system, based on the outcome of descriptive analytics that show the average parking lot occupancy at a specific hour on a specific day is unusually low or high, a smart parking operator might drill down further to understand why that has happened. They might check other data sources to do this – for example, checking the weather and holiday data.

- **Predictive analytics**: This type of analytics focuses on the future – in other words, what will happen based on the historical data we have.

  There are two ways to achieve this type of predictive analytics; the traditional one is by using advanced statistical methods such as logistic regression and clustering, while the other one is by utilizing ML techniques such as supervised and unsupervised learning.

  In a smart parking system, a question such as what the likelihood is of having all parking spaces in a city be fully occupied all day in the next 24 hours, week, or month would be useful.

  As stated earlier, in *Chapter 1, Introduction to the IoT – The Big Picture*, when we talked about the smart parking use case, one of the stakeholders of smart parking solutions is city council administrators. They are usually looking for an outcome from predictive analytics to help them plan ahead – for example, in the aforementioned case, they might add more parking spaces in the city to avoid congestion.

- **Prescriptive analytics**: This type of data analytics focuses on questions such as what actions should be taken based on the outcome of the predictive analytics.

  In the smart parking example, predictive analytics predict full parking occupancy in some cities or areas. Accordingly, what action should a smart parking operator take? Prescriptive analytics might recommend adding additional parking spaces – as explained before – or recommend something else (for example, the occupancy across parking spaces in the city may not be well distributed; hence, there may be no need to add parking spaces and instead distribute the load evenly with other nearby spaces).

Those are the most common data analytics types. In any project, to get the maximum value of data analytics, you must be a subject matter expert in the business domain you are covering or get help from an expert business analyst to list the data analytics questions you are looking for answers to.

Now, in the next section, let's jump into some technical details of how to design and implement data analytics solutions for your IoT solution by exploring different data analytics solution building blocks.

Data analytics solution building blocks

Let's go step by step here to understand the different solution building blocks required to build a data analytics solution.

Data sources

I know this book is about IoT, and we mentioned earlier that IoT is one of the greatest data sources for any business organization, but based on the analytics questions you have to hand, you might need some other data sources to do a proper analysis and answer those business data analytics questions.

For example, you might want data from an external data source such as weather data or an IoT device's metadata, as the data coming from IoT devices is usually raw and small due to bandwidth and power limitations; hence, you can enrich raw data in the cloud by using another IoT device data repository, such as IoT device metadata or IoT device registry data.

To conclude, you might require more data sources besides data coming from IoT devices, so identifying those data sources should be the first step in defining a data analytics solution.

Data ingestion

You need to identify how data will be ingested into the analytics data store; for traditional data sources such as databases and files, ETL tools usually use techniques such as database migration services, transferring files through SFTP protocols, a storage gateway, or even moving data using external mobile hardware storage devices. This works perfectly with data batch processing, meaning transferring and processing raw data from data sources on a daily, weekly, or monthly basis to the central data storage destination. However, for data coming from IoT devices that naturally stream, how will it be ingested into the central data source destination? Note that IoT telemetry data is just one example of streaming data; other examples include website clickstreams, database event streams, financial transactions, social media feeds, and location-tracking events.

Here, we focus on ingesting IoT stream data. As explained in the earlier chapters, IoT devices send IoT data through the MQTT protocol (or any other IoT application protocol) to the IoT message broker in the cloud.

Now, it is the IoT message broker's responsibility to forward the IoT data to a streaming processor component that can deal with it.

To conclude, in an IoT data analytics solution, we will usually have two types of data to deal with (that is, to process data) – data coming in batches and data coming as streaming events. In the next section, let's go through one data processing architecture paradigm that deals with both types of data.

Lambda architecture

Lambda architecture (note that this is not AWS Lambda) is a data processing architecture that is used in big data and advanced data analytics solutions to support historical (that is, traditional) advanced data analytics (ML) and near real-time data analytics.

Let's explore the different components of the architecture to better understand how implementing a conceptual architecture will enable you to build traditional and near real-time data analytics solutions for your large-scale and production-grade IoT project:

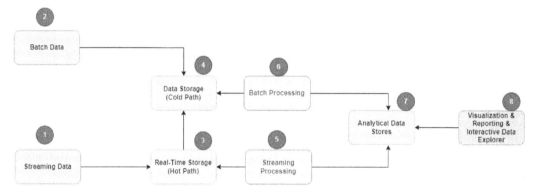

Figure 7.1 – Lambda architecture

In the Lambda architecture, we have the following:

1. **Streaming data**: We have explained before what streaming data is and what the sources of streaming data are, such as IoT devices, users' online activities, and application logs (that is, logs that are treated as events as part of cloud-native best practices).

   This streaming data should be landed first in streaming data storage (hot storage) to be processed immediately or in near real time.

   A copy of the streaming data should be stored in the master data storage (cold storage – the data's single source of truth) for archiving or historical/traditional or batch data analytics purposes; this way, you also secure the durability of the streaming data and it will never be lost, and you can replay again – if needed – the real-time data processing or analytics that you have done in the hot storage in the past.

There is no clear or standard definition for what hot storage and cold storage are, but our view is the following:

- **Hot storage**: This is where data (that is, streaming data) is stored in super-fast storage media. The data is accessed frequently, and the storage cost for the hot storage is expensive due to the characteristics we have just mentioned. Accordingly, data retention in this type of storage is always limited (that is, data is typically stored for 24 hours or 7 days at a maximum); some hot storage solutions offer a longer retention period, but that will ultimately come at a cost.

- **Cold storage**: This is where data (that is, batched and streaming data) is stored in slightly slower storage media compared to hot storage media. The data is usually accessed infrequently or never at all, and the storage cost is cheaper compared to hot storage. Accordingly, data retention is unlimited (that is, you can store data for 10 years or more if you wish, or as per the business or governmental regulations you might have for data retention).

2. **Batch data**: The other type of data that you might need to build your IoT data analytics solution is batch data, which is usually extracted from data source systems periodically (daily or weekly).

   Batch data will usually land in or be loaded into a cold storage layer to be processed further by batch processing.

3. **Real-time or data streaming storage (a data hot path)**: As mentioned earlier, streaming data will land into streaming storage, which is a kind of hot data storage.

   Apache Kafka is a good example of streaming data storage. Although we mention Kafka as an example, there are lots of technological choices in each component of the Lambda architecture. To give an example, if you are using or deploying a data analytics solution into a public cloud, then you will see cloud providers offering their native or managed services for each component of Lambda architecture.

   In AWS, for example, for streaming data storage, you could go with Amazon Kinesis or even a managed Kafka service in AWS (Amazon **Managed Streaming for Apache Kafka** (**MSK**). In Azure and Google, you will also find similar options.

   For IoT data ingestion, the story so far is that IoT data is sent by IoT devices using MQTT or another IoT application protocol, and then it lands into an IoT message broker (that is, a broker that supports MQTT, HTTPS, and so on). The IoT message broker will send the data to a streaming message broker or storage such as Apache Kafka.

You might ask, *Hold on for a second – why don't the IoT devices just send the IoT data directly to the Kafka cluster? Why do I have to first go to the IoT message broker and then Kafka?* This is a very good question – in fact, you can send IoT data directly into the Kafka cluster; no one will stop you. However, it is not recommended and sometimes not feasible either. Why? Let's dig further into that and see.

There are many reasons that prevent Apache Kafka from being the perfect choice for ingesting (initially) IoT telemetry data from IoT devices directly. Here are a few:

- Kafka clients are somehow complex and require more compute resources in IoT devices compared to MQTT client libraries, which require only a low-resource footprint.

- Kafka does not support a large number of topics. In IoT use cases, you might require thousands or even millions of topics; hence, it will be a problem if you use Kafka in such a setup.

- Kafka was built mainly for integrating and moving data quickly between several internal systems. An IoT use case is a bit different, as the number of connected clients will be massive and external (that is, over the internet). So, Kafka is perfect for internal communications and integration between different systems or microservices but not the best option for external integration, especially in IoT cases where there is a massive number of devices/clients that need to connect and send telemetry data.

- The Kafka protocol does not support some important IoT features such as a last will and testament.

Because of the aforementioned reasons and others, we recommend that the pattern should be as follows: IoT device -> IoT message broker -> Streaming message broker.

4. **Streaming processing**: You have captured the raw streaming data in the streaming storage layer (*step 3*); now, it is time to process the streaming data to get value out of it. You will need a streaming processing engine to enable you to process and analyze real-time or near real-time streaming data. We will explain more about streaming processing later in this chapter.

5. **Central data storage (data cold path)**: This is the solution's master data storage or the data's single source of truth. All kinds of data (streaming or batch and in any format) should be stored securely and durably in this storage layer.

This is an important component in any data analytics solution. There are many implementations and views for central data storage; some might use databases, data warehousing, and so on as an implementation for central data storage, but the most modern and best fit for the layer is what is known as the data lake storage layer. We will explain more about the data lake later in this chapter.

6. **Batch processing**: As with traditional data analytics, after you have captured (extracted and loaded) all raw data into a centralized data storage layer, you will need to do some transformation or data processing. Then, you load the processed data into a data store fit for the purpose of analytics. We will explain more about batch processing later in this chapter.

7. **Analytical data stores**: Part of the data processing job in both streaming data processing and batch data processing is to transform and prepare data into a proper format that is super-efficient and well suited for running the required data analytics queries – for example, data in a tabular format is still considered the best format for running analytics queries, especially as it uses the powerful, historical, and well-known query language SQL. You might have noticed that we have called it analytical data stores with an s at the end to highlight that in your data analytics solution, you might require different types of data analytics data stores, as the data consumers might have different requirements. There's no single data analytics data store that is suitable for all analytical consumers. We will explain more about different data analytics data stores later in this chapter.

8. **Visualization, reporting, and an interactive data explorer**: These tools or technologies are important building blocks in your IoT data analytics solution, and without them, you don't really have a proper data analytics solution, even if you have implemented it perfectly in all the other components that we have mentioned. Why? Simply because, ultimately, you will need to show the outcome of this data analytics solution to end users or stakeholders. We will explain more about these tools and their consumers later in this chapter.

This is the Lambda architecture and how you can build your IoT data analytics solution based on it in a nutshell. In the next few sections, we will cover the different components of it in more detail. Let's start with central data storage or the data lake.

Central data storage – the data lake

A data lake is a centralized data storage that helps you store a massive amount of structured, semi-structured, and unstructured data in a more secure and durable way.

In an IoT data analytics solution, you will have many data sources that generate massive and different types of data at different speeds (that is, the three Vs of big data – volume, velocity, and variety), so the data lake is the best-fit data storage for this type of data.

Structured data is databases, spreadsheets, or data that can be represented in a tabular format (rows and columns). Semi-structured data is JSON, XML, and so on. Unstructured data is videos, audio, images, emails, and so on.

Implementation-wise, there are different views on how to implement a data lake. Some might say, *We already have centralized data storage, which is a data warehouse, so you can say the data lake concept is implemented in that way.* Others will say, *We can have a centralized database act as a data lake.* However, we believe that data warehouses and databases are not a true definition or implementation of a data lake; let's dig further to understand why.

Databases or data warehouses are not designed for ingesting raw data of different formats and structures. With databases, you need first to define a schema, and data that will be loaded into the database must follow the database/data warehouse schema. In a data lake, we need first to ingest and store raw data in any data format and structure, and later, when we process such data, we can build the schema. In other words, databases and data warehouses enforce what is known as a schema on write (that is, you build the schema first before data can be ingested or accepted based on the defined schema), while data lakes follow what is called schema on read (you build the schema – after the data has already been ingested – to be able to read the stored data in the data lake in a structured way).

Related to the previous point, data lakes support the ELT model, which means you first load data in its raw format into the data lake (later, we will see how to process this data), while databases support the **ETL model**, which means you need to first transform raw data you receive into a format that is acceptable by databases (that is, a database-defined schema).

Obviously, with big data and IoT data analytics solutions, the ELT model is the preferred one, as it usually has different sources of data and formats, and you just need to ingest such data into a central location instead of transforming each source of data to the database format, which will take time and become a bottleneck for data ingestion.

Another challenge with databases and data warehouses is scalability. Scaling relational databases is limited; they can easily scale vertically (that is, add additional compute resources to the database server, although vertical scalability is limited), but it is not easy to scale them horizontally (that is, add additional servers or nodes to the cluster). We don't need to go into detail or debates here about using database sharding or partitioning to scale them horizontally, and so on, as even with such database sharding or partitioning techniques, horizontal scalability is still a big challenge for relational databases due to the mandatory **Atomicity, Consistency, Isolation, and Durability** (**ACID**) properties that they must follow. Alternatively, data lakes are usually highly scalable and can store (to some extent) an unlimited amount of data.

Okay, now we know that a data lake is a good concept and there is a big difference between it and databases and data warehouses, how do we then build the data lake layer in our IoT data analytics solution?

A data lake is usually implemented using object storage solutions. If it is in the cloud, you will have most of the public cloud providers offering object storage services such as Amazon S3, Azure Blob storage, and so on, which you can use as your data lake layer. Some cloud providers provide a managed data lake solution, as a data lake sometimes becomes very challenging and complex to manage, especially with such a massive amount of data and the different formats it hosts, so it might get out of control. Therefore, you can go with the public cloud-managed option if you want.

If you want to have your data lake on permanently, there are also object storage solutions that can be used on-premises, or you can go with the **Hadoop Distributed File System** (**HDFS**).

So the idea is now simple – for raw data (that is, data coming from IoT devices, social media feeds, batch data, and so on), you need to capture it in a durable, scalable, and performant data lake layer. Later on, you can process the data and transfer it to the required data analytics format (as we will explain later when we talk about batch processing) – for example, you might have BI tools already dependent on a data warehouse for reports and visualization; what you can do is simply transfer the data from the data lake layer to the required format in the data warehouse layer.

You can also transfer that raw data into another format required by other data analytics tools – for example, the data scientist usually requires the dataset in a specific format after applying some data cleaning, preparation, and featuring engineering to run their ML models and so on. This is the beauty of having a data lake as a core component in your analytics solution. As you can see, it serves different needs and is future-proof; this means when other data analytics tools, technologies, or approaches are introduced in the future, they can simply integrate with a data lake (that is, the data's single source of truth) and get whatever data they need to fulfill their data analytics jobs.

A data lake conceptually has different data layers – for example, you can have one layer in the data lake for storing only raw data, another layer for storing preprocessed data, a third layer for storing processed data (or serving data), and so on, so you can organize it as you wish to fit with your requirements. Object storage usually follows a structure similar to a filesystem structure, so you can have one folder called raw data, another folder called preprocessed, and so on.

Data lake – catalog and search

A data lake is a great component in an IoT data analytics solution. It stores all different types of data, as explained, but the main challenge with the data lake is that it stores raw data but has no oversight of the content of that data. To overcome that challenge, usually, besides the data lake there is a catalog service to store metadata about raw data that is stored in the data lake to simplify the search and reach for data, data governance and access control, and data consistency.

For metadata, a NoSQL database is usually used to store the metadata of objects stored in the data lake.

To process or to make SQL-like queries over data lake objects, you end up building another type of repository that holds table definitions for data stored in the data lake, such as the Hive metastore (if you use the HDFS as an on-premises data lake layer) or the AWS Glue Data Catalog (if you use Amazon S3).

A data lake is NoSQL in nature, but everyone likes SQL queries, so in the NoSQL world, you will see different NoSQL solutions that offer **SQL-like queries** to query data stored in a NoSQL service.

In Hadoop, data is stored in the **HDFS**. This represents the data lake part. Data in the HDFS is accessible and processed using the **MapReduce paradigm**, which is a parallel and distributed data processing paradigm. The MapReduce paradigm can be implemented in many programming languages, such as JavaScript and Scala. MapReduce then represents the data processing part – in other words, to access the data stored in the HDFS, you end up writing code that implements the MapReduce paradigm. Some tools such as Apache Hive implement and abstract the MapReduce complexity from the end user and offer a SQL-like query language to access and process the stored data in the HDFS.

Apache Hive is a solution that enables users to interact with data stored in the HDFS by offering an SQL-like query called the **Hive Query Language** (**HiveQL**). You write a HiveQL (similar to SQL) query, and behind the scenes, Hive will transfer it into MapReduce programs and return the result to the HiveQL query. To do this, Hive depends on an essential component, which is the Hive metastore/catalog.

With that, we have explained the data lake and the idea of how to process data in it by exploring the concept of the data catalog. Now, let's move on to another aspect of Lambda architecture, which is data processing.

Now, with data stored in the data lake and with a capability such as the data lake catalog to search and access that data, let's move on to the data processing part and start – since we are here in the data lake section – looking at batch data processing.

Batch data processing

Data (batch and streaming) stored in a data lake in its raw format might be stored in the JSON, XML, or CSV format. A destination analytics system such as a data warehouse or ML will usually require some data validation, enrichment, preprocessing, and data feature engineering (that is, related to ML), applied to raw data in a data lake before loading it into the relevant analytics data store.

There are many batch processing frameworks and solutions to process data stored in a data lake and generate the required outcome expected by different data analytics systems, such as Apache Spark, MapReduce, and Apache Flink. If you are using a public cloud, such as AWS, then you can use solutions such as AWS Glue (which provides managed Spark services – that is, you just submit Spark jobs to AWS Glue that it will run in a serverless way, so you don't need to worry about the infrastructure required for running an Apache Spark cluster), Amazon EMR (which is a managed Hadoop distribution in the AWS cloud that helps you run MapReduce jobs in the cloud without worrying about managing a Hadoop cluster infrastructure, which is usually not an easy task to manage), and the AWS Batch service. Microsoft Azure and Google Cloud also have similar products and services for batch processing.

One important point to note here is that the main objective of batch processing is converting raw data into a proper format for analytics purposes. For data analytics data stores, the best data formats are **Parquet** and **Avro**; they are in the columnar data format, which is perfectly recommended for any data analytics query engine. They also save space in the underlying data storage, which could help in cost reduction as well.

One of the good things about a data lake architecture is decoupling storage from computing. Raw data is stored in a data lake, and when it comes to computing to process it, there are different batch processing frameworks that simply work on top of the data lake and scale separately if needed, as discussed.

To conclude this topic, batch processing starts to work on top of a source data store such as a data lake and then produces output that fits with target analytics data stores. Batch processing can start based on a schedule or event, and it can be controlled by an orchestration tool that triggers, orchestrates, and controls batch processing jobs – for example, an orchestration engine can start a Hadoop cluster to do the batch processing and tear down the cluster after completing the processing.

Explaining different batch processing frameworks is beyond the scope of this book, but as an IoT solution architect, you should be aware of the difference between batch processing and streaming processing and when you should use each one of them or both in your IoT data analytics solution. Let's now cover streaming data processing in the next section.

Streaming data processing

Streaming data including IoT telemetry data, according to the Lambda architecture, is processed first in the streaming data processing storage layer (such as Apache Kafka) and then a streaming processing engine (such as Apache Kafka Streams or Apache Spark).

This enables you to apply actions immediately to the received streaming data, enrich the received streaming raw data, or buffer the streaming data into a small time window and apply near real-time data analytics to it.

To give an example, let's return to the smart parking solution. Here, you can have some business requirements to apply near real-time data analytics and alerts, such as drivers being able to use or reserve only two parking lots in the same parking space within a 10-minute window in a day. If that is exceeded, then additional charges or another business action could be applied.

To achieve the aforementioned use case, you need to store the parking lots' IoT data in a hot storage service and apply near real-time data analytics in a 10-minute window (or fewer if needed). Batch processing will not help here as batch processing usually runs once a day, every 12 hours, or even every 1 hour (still, it can't cover the 10 minutes or fewer requirement), so running queries over the 10-minute window can be applied only using streaming data storage and a streaming data processing engine.

Streaming data processing requirements might be very simple – for example, if the temperature sent from the IoT device is more than 40 degrees, then you need to send an alert (that is, an email, SMS, or calling an API) to the operations team. This very simple data processing can be done in a streaming processing engine, while complex real-time processing can also be done in a streaming processing engine.

Another interesting thing that you can do in a streaming processing engine is inferencing or invoking a pre-trained and ready-to-use ML model in near real time. You simply call the ML model endpoint or API and pass to it the received streaming data; the ML model will evaluate the incoming streaming data and return the result immediately in real time.

The ML model can be a binary classification model where you pass some input data and the ML will reply with yes or no – for example, in the smart parking solution, you could use ML in identifying the parking lot occupancy. So typically, you train your model on the cloud using lots of images that show when the car parking lot is occupied or not, and once the ML model is ready, you simply expose it as an API that receives images as input to evaluate whether the parking lot (that is, what is shown in the received image) is occupied or not. The IoT parking camera device that is deployed in the parking space area will keep streaming the videos/images of parking lots to the IoT cloud core, and then the streaming data will reach the end of the streaming engine, which will invoke the ML API to detect whether the parking lot is occupied or not immediately. The ML model can be deployed in the cloud or the edge.

There are many streaming processing engine solutions on the market. If you are building a data analytics solution on-premises, then you could go with **Apache Kafka Streams** (as the name suggests, it is well integrated with Apache Kafka, which is used as data streaming storage), **Apache Spark** (this the most widely used streaming processing engine in the market at the moment), or **Apache Flink**. If you are building on the public cloud, then most of the public cloud providers have their own streaming processing solutions or managed streaming solutions (such as managed Kafka and managed Spark). In AWS, for example, you have **Amazon Kinesis** (an Amazon proprietary solution) or **Amazon MSK**, and you can use **Amazon Kinesis Data Analytics**, which is well integrated with Amazon Kinesis, to do near real-time analytics using standard SQL in a serverless way. Alternatively, Microsoft offers **Azure Stream Analytics**.

In streaming data analytics, there is one important concept that you should understand: generally, to make a query (aggregating, grouping, and so on), you need a bounded dataset to run that query over. The challenge with streaming data is that, by nature, it is not bounded; it continuously comes and ingests into streaming data storage. So, how will you run your data analytics queries over this continuous data streaming?

To answer that question, streaming data events that are stored in streaming data storage are divided into groups of events called **windowing**. Windowing or event grouping can be done based on any attributes of the event data. However, the most commonly used attribute for windowing is time. So basically, you assign each streaming event a window; by doing this, you can run queries on the window as the data streams now somehow become bounded (for example, a window of 1 minute or 5 minutes). There are four types of window methods – tumbling window, hopping window, sliding window, and session window – but an explanation of these window types is beyond the scope of this book.

Finally, and to summarize this section with an example, the typical flow in AWS IoT implementation will be something such as an IoT device sending telemetry data to AWS IoT Core. Then, a rule configured in the AWS IoT Rules Engine listens to a specific topic where the telemetry data is coming from. The rule will send the data to a streaming processor engine such as Amazon Kinesis Data Streams and also to an S3 data lake (remember the Lambda architecture). Amazon Kinesis Data Streams will do the required streaming processing you might have (that is, enrich the incoming streaming data, do simple logic, fire a Lambda function, and send a topic for an alert to Amazon **Simple Notification Service** (**SNS**)). Then, if you have near real-time data analytics requirements, you can have Amazon Kinesis Data Streams stream the data further to *Amazon Kinesis Data Analytics* to do the near real-time data analytics required. Ultimately, the outcome of the near real-time data analytics can be used for building real-time dashboards or can be stored in a data warehouse for further complex and historical analytics – that is, you store the 10-minute window analytics outcome every day in a data warehouse to be used later for a more complex data analytics purpose (that is, analyze the 10-minute window of sales every day in every month in the last year, or something similar).

With this, we conclude the streaming data processing part, and again, all the concepts explained in this chapter, including data streaming processing and near real-time data analytics, are explained at a very high level. However, if further information is required about these concepts, then we advise you (or the IoT solution architect or designer) to dig further into other relevant books and resources that cover these concepts and architectures in detail.

Now that the batch and streaming data has been processed or completed, transformed, and loaded into data analytics-fit data stores, let's explore in the next section the different data analytics data stores and query engines that we can use in the IoT data analytics solution.

Query engines and data analytics data stores

We learned how to ingest data into hot (streaming) and cold (batch) storage; we learned also how to process both types of data.

The output of data processing, whether batch processing or processing, is then loaded into different data stores for different data analytics purposes – for example, processed data can be loaded into the following:

- **A NoSQL database** (for example, Cassandra, Amazon DynamoDB, Cosmos DB, and Google Bigtable), which can be used as a backend for a real-time dashboard or applications.

- **A data warehouse** (for example, AWS Redshift, Snowflake, Google BigQuery, Azure Synapse Analytics, and Oracle), which is a typical backend data analytics data store that is used in many organizations (and will continue to be used) for building reports, dashboards, and KPIs.

  A SQL or relational-based data store is the best-fit analytics data store due to its solid relational schema, making searches and queries so powerful and efficient in comparison to other data models. A data warehouse takes that concept even further by introducing other types of data schemas (for example, star, snowflake, and galaxy) and other analytics features that enable more efficient queries and reporting.

- **A text-based search engine** (for example, Elasticsearch, OpenSearch, Apache Solr, and Kusto). You might use something such as Elasticsearch for data analytics purposes, so the streaming processor will stream incoming data after applying some processing – if needed – into a data store such as Elasticsearch to be used later by other visualization tools such as Kibana for analytics and reporting.

- **A data lake** (for example, Amazon S3 and the HDFS). Yes, you read it right – a data lake can be used as central storage for raw data and also for storing preprocessed and processed data that can be used to do some data analytics jobs.

  Mainly data scientist teams use the processed data in a data lake for building, training, and evaluating different ML models that they are building.

- **Relational databases** (for example, Amazon RDS, PostgreSQL, and MySQL). You might just use a traditional database for building some simple business reports and dashboards.

- **Query engines** (for example, Athena, Presto, Hive, and Redshift Spectrum). As mentioned before, one of the good characteristics of a data lake is decoupling storage from computing, so data is stored securely and durably in a data lake. Then, if you want to process the data, you launch a cluster for batch processing to process the data that is stored at rest without moving the data to that cluster, as moving such large amounts of data might cause a bottleneck in the network throughput.

  There are many query engines that can process data stored in a data lake on the fly. The concept behind all the query engines is simple; they first build a data catalog of data stored in the data lake and then they offer a SQL-like query to the data consumer to explore such data. Behind the scenes, those query engines transform the SQL-like query to the underlying data processing framework, as explained earlier when we discussed Hive and HiveQL.

In the next section, let's see the different consumers of data analytics stores and query engines.

Visualization and data exploration

Congratulations! Now, you have collected or ingested data (that is, batch and streaming) from different data sources that you have identified for your data analytics solution. You've built near real-time analytics applications with the incoming streaming data, and you have stored the outcome of this near real-time analytics application in a proper data store that you will use to run analytics queries – for example, the outcome of a 10-minute analysis of parking lot occupancy is stored in a table in a SQL database called 10minoccupancy, or you store it in a key-value or document NoSQL database. You have also populated the data warehouse with processed data from the batch processing that you did. Now, it is the time to expose it to the business' end users or data analytics consumers, so you need to visualize the analytics data in a proper format such as charts, maps, and graphs:

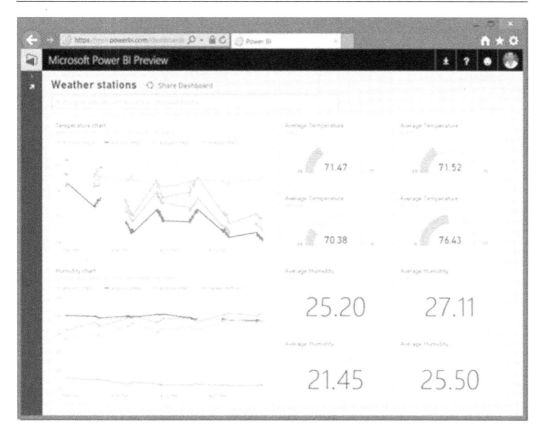

Figure 7.2 – Microsoft Power BI – monitor IoT sensors

So, broadly, we can say that the result of the data analytics can be either visualized for business users or explored (interactively) by data scientists for building different ML models.

There are many visualization tools that you can use in your data analytics solution, such as Microsoft Power BI, Qlik, Tableau, Amazon QuickSight, and MicroStrategy. Those tools are ready-made visualization tools that help you quickly build business reports and visualizations.

You can also use different JavaScript libraries that produce dynamic and interactive data visualizations in web browsers, such as D3.js, Chart.js, and others.

Data scientists often explore the available datasets while building and training ML models; this is normally an interactive and iterative process and requires tools that can query the data quickly across massive amounts of datasets.

There are two categories of tools used by data scientists for data exploration:

- Tools that use or offer a SQL query engine style to interact with the dataset such as Spark SQL (using its native interactive shell user interface or another user interface such as Apache Zeppelin), Apache HIVE (using its native interactive shell or **Apache Hadoop User Experience (HUE)**), or Presto (using its CLI or Airpal tool).

- The other style used by data scientists is a programmatic style, meaning they are dependent on programming languages to do data exploration such as R (user interfaces such as **RStudio** and the R interactive shell), Python, and Scala (user interface tools supporting these include Apache Zeppelin, Jupyter Notebook, and the Spark interactive shell).

Now, let's summarize this section and add some additional design considerations in the following points:

- Always remember that the goal for all IoT technologies covered in this book and other books is to bring data out of the physical world to get more insights and to act upon those insights to ultimately offer better services to your customers, increase revenue, and reduce operational costs.

- IoT-generated data is classified as streaming data; therefore, you should have in your IoT architecture a very solid stream message broker and stream message processor.

- When it comes to IoT data analytics or big data analytics, in general, always follow well-known architecture paradigms such as Lambda and/or kappa architecture, which enable you to deliver near real-time data analytics, alerting, and notifications, as well as traditional data analytics.

- A data lake is the backbone of a modern data analytics platform. There are many benefits to having a data lake in your IoT solution ecosystems and architecture. It can grow indefinitely (a data lake decouples storage from computing), it is future-proof, as it integrates with many different processing and data analytics tools, and finally, it is a low-cost storage service compared to other storage options.

  If you were to ask where you should build a modern data analytics architecture (on a public cloud or a private cloud/on-premises), our recommendation would be to build it on the public cloud, if applicable. In the public cloud, you will have the scale (that is, massive compute resources) and lots of managed services available for each layer of your IoT data analytics architecture – for example, Amazon S3 is an object storage service that works perfectly as a data lake layer. AWS offers a managed data lake solution to reduce some of the operational burdens of managing a data lake yourself. In batch processing, there is a huge set of managed services provided by the public cloud provider; it is the same with stream processing and other services required in a complete IoT data analytics solution.

ML and AI solutions in the public cloud are a big advantage as well in deploying an IoT data analytics solution in the public cloud. Most public cloud providers provide managed solutions and services for building, training, and inferencing ML models, besides a huge set of ready-made and pre-trained AI and ML services such as video and image analytics, fraud detection, demand forecasting, real-time translation, text to speech, transcription, and many more services that are ready to be consumed.

- Yes, we recommend building the data analytics platform on the public cloud, but we know that in large enterprises, that decision might not be acceptable, especially when we are talking about customer data, so there may be some concerns about hosting customer data outside organizational boundaries. There are solutions and mitigation plans to overcome those concerns such as using data anonymization, data masking, or data encryption, but some enterprises or regulations still don't allow customer data to be stored in the public cloud. Because of this concern, we focused more in this chapter on concepts. So, if you want to build a data lake on-premises, then you can build it using the HDFS, for example; if you want batch processing, you can go with Spark or other open source solutions that are available.

- So, if needed, you can build your data analytics platform on-premises, but it will not be an easy journey and not as efficient as being built on the public cloud.

- Now, let's assume we will go with the option of a data analytics platform on the public cloud. What options do we have? The answer is we have two options: we can build that architecture ourselves using the different services that are provided by the public cloud (or bring some open source or commercial tools and technologies to the public cloud IaaS and manage it ourselves), or we can use a fully managed IoT data analytics service that some public clouds offer (in the next section, we will talk about the AWS IoT Analytics service). So, which option should we choose?

  Our recommendation is to use the fully managed IoT analytics service if you are trying to do a quick proof of concept, launch a quick IoT data analytics product, or if the data processing logic is not too complex, but if you want more control and customization, then you should go with the option of building a data analytics platform on the cloud using different cloud provider services or open source services on a public cloud IaaS.

To continue our journey with the AWS IoT platform and in relation to the last point mentioned in the previous section, in the next section, let's talk about another service provided by the AWS IoT platform covering IoT data analytics – the **AWS IoT Analytics** service.

AWS IoT analytics

AWS IoT Analytics is a fully managed IoT analytics service that helps you collect, preprocess, enrich, store, and analyze IoT data at scale.

The AWS IoT Analytics service is fully integrated with AWS IoT Core; this is a great feature, as you can build IoT data analytics solutions more quickly using this managed service.

The AWS IoT Analytics service also integrates with other AWS and non-AWS services, and it accepts data from many data sources, so you might use a different IoT core from a different vendor and use the AWS IoT Analytics service for analytics purposes. In other words, you don't need to use AWS IoT Core to use AWS IoT Analytics; however, if you already use AWS IoT Core, then it is much easier to go with AWS IoT Analytics, as it is fully integrated with AWS IoT Core.

Let's see AWS IoT Analytics' main solution building blocks:

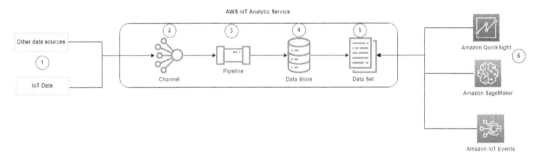

Figure 7.3 – The AWS IoT Analytics service

The architecture is simple, as per the following:

1. **Data sources**: There are two methods to ingest the data into AWS IoT analytics, first through AWS IoT Core. AWS IoT Core is integrated with the AWS IoT Analytics service through the AWS IoT Rules Engine, which flows like an IoT device published message or IoT data to AWS IoT Core. There is a rule configured in the AWS IoT Rules Engine that listens to the topic messages are sent to. The rule is configured to forward the data to the AWS IoT Analytics channel (see the **channel** component discussed next).

    The second method is by using the AWS IoT Analytics APIs to send data directly, such as the `BatchPutMessage` API.

2. **Channel**: In AWS IoT Analytics, the raw data is ingested into a channel, which stores the data it receives in S3 buckets. The channel S3 bucket can be a customer-managed or an AWS-managed bucket.

You can create one or more channels if you want, but you should be careful about the cost of storage.

3. **Pipeline**: This is where data processing is done. The pipeline consumes messages from the channel, and data processing is done through a series of data processing activities. Activities are things such as data transformation and enrichment, mathematical transformation to normalize device data, or invoking an AWS Lambda function for advanced and complex data transformation logic if needed.

   You can create one or more pipelines if you want; you can have one or more pipelines per channel.

   In the pipeline, you define the schema of how raw data in the channel is represented. You define the attributes of the messages, either manually or through a JSON template. Defining message attributes or schemas will help you when you do transformation or data processing activities such as removing attributes from a message, adding attributes to a message, calculating a message attribute, appending device registry info, and appending device shadow info.

   The processed data from the pipeline is ultimately sent to the AWS IoT Analytics data store (see the **data store** component discussed next).

   To give an example, you can have a pipeline activity to fill some missing data, perform common calculations such as Celsius-into-Fahrenheit conversion, or enrich data with an external data source such as a weather forecast.

4. **Data store**: This is where the data processed by the pipeline is stored and managed. You can have multiple data stores; you can have one data store per pipeline or one data store as a destination for many pipelines.

   A data store is also backed by Amazon S3 buckets, which can be managed by a customer or managed by AWS.

   The processed data format is important, as you will run analytics queries over data stores to retrieve the data you are looking for; hence, a proper format should be selected for that purpose. At the time of writing, AWS IoT Analytics currently supports JSON and Parquet file formats (the columnar storage format is always preferable, as it is efficient in storing and querying large volumes of data). The default file format is JSON. You can specify only one format when creating a data store, and you can't change the file format after you have created the data store.

5. **Datasets**: The data analytics queries run over datasets, not data stores. AWS creates that layer (that is, datasets) in the middle, so you can define what data you want to query. It is similar to the view concept in the relational database, so you can create a dataset by selecting all data from the data store, which eventually means you will access the data store directly, since you get all data from the data store, or you can create datasets based on queries and filters that retrieve subsets of data stored in the data stores.

    The benefit of having this layer (that is, datasets) is when you have many consumers of the data, you can create a dataset for each relevant data consumer with the relevant set of data.

    Datasets are also backed by Amazon S3. You query the dataset by using a SQL-like query. We have explained before in this chapter how you can run SQL-like queries over a non-relational database or store.

    There are two types of datasets:

    - **SQL dataset**: This is a kind of materialized view of the data store, as explained earlier (that is, you define a SQL query and filters over data stored in the data store; the outcome of the query is populating the SQL dataset). This type of dataset is typically used for reporting purposes and dashboards; you can also have notebooks, such as Jupyter notebooks, to interactively explore the data in that type of dataset. To refresh a view or dataset, you define a schedule to refresh data stored in that dataset. You can also define the data retention period for that dataset. The data of the dataset is stored in Amazon S3, or it can be sent to the AWS IoT Events service.

    - **Container dataset**: This type of dataset is created by either copying a SQL dataset or linking to a SQL dataset; with this, you have what is known as a container dataset, and then you configure a container image, containing the application analytics logic that you need to run over that container dataset. You define the container image as usual by defining things such as the required compute resources for that container (that is, a vCPU and memory), the container image, and some input variables (note that the value of one of those input variables is usually referring to the container dataset URI; remember the dataset stored in the S3 bucket at the end so that you can easily get the URI of the dataset) and output variables (note that one of the output variables values will usually be referring to the URI where the result of the data analytics running on that container dataset will be stored). You can run some anomaly detection analytics or any other type of analytics every 15 minutes or so – so, for example, the container dataset will be created every 15 minutes, and a containerized application containing the analytics logic will run that anomaly detection on 15 minutes' worth of data and put the result or the output file into an S3 bucket defined in the output variable for that container dataset.

6. Finally, and as usual in any complete or E2E data analytics solution, we come to the visualization or consumption part of the analyzed datasets. You can use ready-made visualization tools such as Amazon QuickSight to build dashboards or reports on top of available datasets you've created in the previous step, or you can interact with datasets using data science tools such as notebooks to explore data to build and train an ML model.

    You can launch a Jupyter Notebook instance in the AWS SageMaker service and connect that Jupyter notebook to the IoT Analytics dataset and start working with that dataset.

The AWS IoT Analytics service enables you to build a quick IoT data analytics solution; interestingly, there are some other AWS services that are somehow related or usually used or combined with the AWS IoT Analytics services and E2E IoT data analytics solution architecture in AWS. These services are AWS IoT Events and AWS IoT SiteWise.

In the next section, let's briefly explain these services.

Other IoT analytics-related AWS services

The AWS IoT platform offers some services that cover business and technical areas needed for building a complete E2E IoT solution in AWS. In the next section, we will start with the AWS IoT Events service.

AWS IoT events

AWS IoT Events is a fully managed service that detects and responds to events from different IoT sensors, services, and applications.

You can say that AWS IoT Events is a fully managed implementation of a **Complex Event Processing** (**CEP**) architecture where the CEP engine aggregates, processes, and analyzes a massive stream of data or events to gain real-time insights from the events that occurred.

The definition of event here means a business event, such as a failure in equipment or fraud detection.

So, business domain experts can build a detection model for their business using the AWS IoT Events console by defining the **input** that the detection model needs to work upon, the **logic** that they need to apply on the input using simple if-then-else statements, and finally, the **action** that will trigger once the business event or condition is detected. Actions can be as simple as an alert or notification sent to an end user or a more complex custom action; for complex actions, you can invoke an AWS Lambda function to perform the logic.

How is the AWS IoT Events service related to the AWS IoT Analytics service? As explained earlier, the AWS IoT Events service requires input or data source input. One of those inputs is AWS IoT Analytics, which integrates very well with AWS IoT Events. AWS IoT Analytics sends to AWS IoT Events processed data, so the AWS IoT Events detector model can work on top of it.

The other input to the AWS IoT Events service can be the raw telemetry data coming directly from the IoT device (you can configure an IoT rule to forward the IoT data going to IoT Core to be sent to AWS IoT Events). You can also ingest data into AWS IoT Events from third-party applications via the AWS IoT Events direct ingest API (the Put_ Signals API).

Within IoT Events, you can define the event detection model in one of two ways. The first way is by using the AWS IoT Events console to define the conditions under which an event occurs and trigger actions when the conditions are evaluated to true. The second option is to programmatically create an event detection by calling the Create_Detector API.

The detector model is similar to building a state machine for your equipment (product), service, or business process – for example, the initial state of your connected product is normal; then, if the temperature is greater than X and humidity is greater than Y, you move the state of your connected product to a NotNormal state, and so on.

To summarize, the flow or pattern is usually as follows – IoT data is cleaned, enriched, processed, and stored in a dataset, and then that dataset is sent to AWS IoT Events, where the business detection model is built and ready to be used. Then, if there are business events detected, AWS IoT Events can trigger an action in near real time. We explained before that the dataset can be scheduled to be populated from a data store, so you can have that process automated by running every 5 minutes or so, for example.

AWS IoT SiteWise

AWS IoT SiteWise is a managed service that lets you collect, model, analyze, and visualize data from industrial equipment at scale. With AWS IoT SiteWise Monitor, you can quickly create web applications for non-technical users to view and analyze your industrial data in real time. You can gain insights into your industrial operations by configuring and monitoring metrics such as mean time between failures and **Overall Equipment Effectiveness** (**OEE**). With AWS IoT SiteWise Edge, you can view and process your data on your local devices.

AWS IoT SiteWise usually comes into the picture when we are dealing with IoT industrial customer solutions. These customers are always looking for a secure, cost-effective, and reliable solution to ingest all data generated on hundreds of industrial sites that have tens of thousands of **Programmable Logic Controllers** (**PLCs**) and sensors in the cloud IoT Core. This enables them to monitor the industrial devices, processes, and equipment in near real time, and also to visualize the key measurements and metrics of those devices, processes, and equipment in near real time.

Usually, there is a massive number of sensors and PLCs in the industrial sites, and they use the **OPC-UA** protocol or other industrial protocols (**Modbus** and **EtherNet/IP**) in communication.

The way AWS IoT SiteWise works is that, on the customer side or on the industrial site, you deploy a gateway to interact with the OPC-UA (or the other industrial protocols such as Modbus and EtherNet/IP) server deployed on the site. The AWS IoT SiteWise connector runs on the common industrial gateway devices running AWS IoT Greengrass and reads data directly from servers and historians over the OPC-UA protocol (or other industrial protocols).

The AWS SiteWise service has two parts: one part (the AWS SiteWise connector) is deployed on the edge or in the industrial site, and the other part is the AWS SiteWise cloud service that receives the ingested data coming from the AWS SiteWise connector.

The AWS SiteWise service enables operation engineers and site managers to build what is known as asset modeling, which is, in short, a virtual representation of the industrial site assets. Once an asset model is built, a dashboard can be created for the operations and process engineers to monitor the factory or industrial site assets in near real time.

What is the relationship between AWS IoT SiteWise and AWS IoT Analytics? AWS IoT SiteWise is the entry point for raw data coming from a factory or industrial site, so it will send the data to AWS IoT Analytics for processing via the AWS IoT Rules Engine (or directly using the AWS IoT Analytics ingest API) as explained earlier. In other words, AWS IoT SiteWise is one of the data sources for the AWS IoT Analytics service.

With this, we come to the end of the section on the AWS IoT platform services, where we focused more on the design part and how the services are working together at a high level. As we know, there is already a massive amount of online documentation, tutorials, workshops, and so on that show you how to use the services step by step.

We don't want to also give step-by-step examples, as we just used the AWS IoT platform as an example. You might already have another IoT platform from another vendor or might even intend to build an IoT platform from scratch.

In the last chapter of this book, we will go through an E2E industrial IoT reference architecture using the AWS IoT platform to recap what has been explained in this book in general, connecting the dots of the AWS IoT platform services and other AWS services, and how they work together.

Summary

In this chapter, you learned about IoT data analytics, and the true and ultimate business benefits of IoT technologies, which is getting data from the physical world to gain more insights and act upon that data. You also learned about traditional data analytics versus modern or advanced data analytics, and the different types of data analytics (descriptive, diagnostic, predictive, and prescriptive).

You also learned about the different building blocks of IoT data analytics solutions such as Lambda architecture, data lakes, batch and streaming data processing, query engines, different analytics data stores, visualization, data exploration tools, and technologies used in data analytics solutions.

You learned about the fully managed AWS IoT Analytics service, and how it works and integrates with other related AWS services such as AWS IoT Events and AWS IoT SiteWise.

In the next chapter, we will cover some of the architecture and design paradigms and concepts used in building modern applications such as microservices, API gateways, event-driven architecture, and so on. These architecture and design paradigms are very important for an IoT solution architect to be familiar with, as they are considered the core components in any production-grade and large-scale IoT solution, as we will explain.

Section 3:
IoT Application Architecture Paradigms and IoT Operational Excellence

The objective of Section 3 is to help you understand the different modern application architecture paradigms that are used in building large-scale IoT backend layers. You will also learn about best practices for how to operate and efficiently run large-scale and production-grade IoT solutions. In the last chapter of this section, you will get some insights into the future of IoT and the emerging technologies that will impact its future.

This part of the book comprises the following chapters:

- *Chapter 8, IoT Application Architecture Paradigms*
- *Chapter 9, Operational Excellence Pillars for Production-Grade IoT Solutions*
- *Chapter 10, Wrapping Up and Final Thoughts*

8
IoT Application Architecture Paradigms

In large-scale and production-grade IoT solutions, you will find that most of the architecture, design, and development discussions, workshops, and efforts are focused on the IoT cloud or the IoT backend cloud and its different components and solutions that ultimately serve to deliver fully fledged IoT solutions and products to customers and end users.

Earlier in the book, we covered the IoT cloud and the IoT backend cloud in some detail. We looked at the infrastructure, solutions, and software components of the IoT cloud, although, to some extent, we focused on the kind known as the serverless IoT cloud, where most of the IoT cloud services and solutions are provided out of the box by the IoT cloud provider or vendor. However, we did mention the option of building IoT backend cloud solutions from scratch if needed. In this chapter, we will deep dive into some application architectures and design paradigms that will help you as an IoT solution architect in designing and building IoT backend cloud components and solutions to meet your needs.

The **application architecture paradigms** that we will cover in this chapter can be used in designing and building IoT cloud or edge solutions, so can be considered generic application architectures and design paradigms.

We will use the digital transformation evaluation roadmap or the modern cloud-native app architecture to explain some important architecture paradigms used in building IoT backend components and solutions. We will not spend much time on the history of different application architecture paradigms, as we would rather go into the current hot topics that are causing a lot of debate within the different software communities and our recommendations around them, which will be based on our day-to-day work experience.

So, for example, for the topic of designing and delivering modern app architectures for the IoT backend, we will not explain in detail the history of monolithic architecture, **Service-Oriented Architecture (SOA)**, N-tier architecture, and so on. Instead, we will focus on the microservice architecture, which is now well known and has become famous as the backbone architecture for building modern apps.

Microservices are just one example of what we will explain in this chapter. We will examine other paradigms and design concepts that work closely with microservice architecture, including API gateways and service meshes.

Fundamentally, microservices, containers, **Continuous Integration/Continuous Deployment or Delivery (CI/CD)**, and automation are considered the key pillars of cloud-native modern app architecture. In other words, these technologies and architecture paradigms are key enablers for delivering truly modern cloud-native apps. So, we will start by first defining the concept of cloud-native and then will move on to microservices and other concepts.

In this chapter, we will cover the following topics:

- Cloud-native versus cloud-ready – what is the difference?
- The Twelve-Factor App methodology
- Microservice architecture
- IoT application reference architecture

Cloud-native versus cloud-ready – what is the difference?

Lots of people get confused between cloud-native software and cloud-ready software, so let's start by understanding what cloud-ready means. Traditional software running on virtual machines with no requirement for any specific hardware architecture can be migrated to a public or private cloud easily. This software can therefore be classified as cloud-ready software. The type of migration here is called shift and lift and is always the number-one choice for cloud migration projects.

Now let's go into what cloud-native software means. In short, cloud-native software or applications are those designed and built from the beginning to take advantage of the cloud computing model. In other words, they are born on the cloud from day one. Still not clear? No problem, let's dive deeper.

Let's take one aspect of software development practice, logging, as an example to better understand the difference between cloud-native and traditional software. In traditional software, we used to write or append application or software logs into a file on the node where the software was running. Then, another background process would run to extract the log files generated in the local node and send them (either at the end of the day or at whatever time you configured it) to a centralized logging system, such as Splunk. Then, the operation support team used that logging and monitoring system to evaluate the log files and detect trends, problems, or other business value that could be gained from them.

I hear you saying, "*I can write the application logs to an external log file in another shared (in other words, centralized) node in the network instead of writing locally*". This is not a recommended approach, as it might kill your application performance. Logging is typically an extensive process in terms of input and output operations on the storage layer, which will ultimately impact the performance of your application.

Logging in cloud-native software is different. In cloud-native software, you expect the node running the software to go down at any moment, and another node to come to life to replace that stopped or terminated node. Or, you might have an autoscaling group that continuously adds and removes nodes dynamically (based on CPU or memory utilization, or custom metrics). In that case, if you use logging as in traditional software, then you might lose some logs when nodes are terminated or shut down before the background process that runs on those nodes has sent the local logs to the centralized logging system. This will be a scary situation, because if you encounter a problem, you won't have any clue why that problem has occurred.

Cloud-native software apps are usually designed to be 100% stateless because they will be deployed in the cloud, which is elastic (that is, nodes go up and down in minutes or less). Therefore, something such as logging should be designed in a specific way to fit the elastic nature of the cloud. In cloud-native software, logs are treated as event streams, meaning the software component will emit log events to a centralized logging system if something goes wrong or for information purposes (that is, for built-in metrics or KPIs for the software app).

Treating logs as event streams is one of the principles of the very famous cloud-native methodology called *Twelve-Factor App*. We will briefly go through the 12 factors shortly to better understand what is meant by designing and building cloud-native software.

The logging example we mentioned shows the difference between cloud-ready and cloud-native software. Traditional software might be cloud-ready, but to be cloud-native, it might need to be revisited/updated in terms of design and source code.

In cloud migration and transformation projects, transforming traditional software into cloud-native software is the best migration option. However, it is also the most complex and costly, especially if the software to be transformed is very old and complex.

Before we conclude this section, let's see the official definition of cloud-native from the Cloud Native Computing Foundation:

> *Cloud-native technologies empower organizations to build and run scalable applications in modern, dynamic environments such as public, private, and hybrid clouds. Containers, service meshes, microservices, immutable infrastructure, and declarative APIs exemplify this approach.*
>
> *These techniques enable loosely coupled systems that are resilient, manageable, and observable. Combined with robust automation, they allow engineers to make high-impact changes frequently and predictably with minimal toil.*
>
> *The Cloud Native Computing Foundation seeks to drive the adoption of this paradigm by fostering and sustaining an ecosystem of opensource, vendor-neutral projects. We democratize state-of-the-art patterns to make these innovations accessible for everyone.*

This definition perfectly shows the different pillars used in building cloud-native solutions, such as microservices, containers, and APIs. This aligns with the topics we will be covering in this chapter, but before we jump into microservice architecture, let's briefly cover the Twelve-Factor App methodology in the next section.

The Twelve-Factor App methodology

The Twelve-Factor App methodology is a widely accepted methodology that defines a set of principles and best practices that software developers should follow to design and build modern cloud-native applications. In essence, cloud-native applications should be portable, scalable, and fast to deploy.

The following is a quick examination of the factors that make up the Twelve-Factor App methodology:

1. **Code base**: *Have one code base tracked in revision control, and many deploys.* This factor mandates having a single code base or source code repository for each service or microservice of your application. In GitHub, GitLab, Bitbucket, or whichever source code repository you use in your project, you should create a dedicated repository for each microservice. This gives you flexibility and full control over each microservice, so the team assigned to build that microservice can do any updates or deployment without impacting other teams working with other microservices.

2. **Dependencies**: *Explicitly declare and isolate dependencies.* This factor mandates that each microservice isolates and packages its own dependencies. This is similar to the goal of principle 1 (code base) – both embrace changes in a service or microservice without impacting other services in the project.

3. **Config**: *Store config in the environment.* This factor mandates that the configuration needed by the application should be stored externally and not inside the application itself. This is a very important principle as it gives you much more flexibility since the software will run in any environment without changing the software itself. You just provide the configuration of the deployment environment as an input (through environment variables or any other mechanism) to the application and the application will react and work accordingly. For example, software running in a development or integration environment will use some testing or development databases or some other testing/development systems. For this, you can provide the dev configuration file as input for the software deployed in the development environment. The dev configuration file will contain the configuration of the dev/testing databases and other dev/testing systems. Likewise, in production, a production configuration file should be passed as input to the same software, referring to production databases and systems, and so on.

4. **Backing services**: *Treat backing services as attached resources.* This factor mandates that backing services, including different data stores (relational databases, NoSQL databases, caches, object storage, and so on), message brokers, identity services, streaming services, and monitoring and analytic services, should be exposed to the cloud-native software component via URLs. For example, a microservice that needs to integrate with a caching service will have the URL and other integration and access details of that caching service configured in the microservice's external configuration file or system.

    With this factor, you decouple the core software component from other helpful or ancillary resources (such as backing services). For example, say your software component interacts with the AWS RDS for PostgreSQL service, but you have been asked to move to another database service, such as Azure Database for PostgreSQL. You simply change the URL of the backing service (in other words, from the RDS for PostgreSQL endpoint to the Azure Database endpoint) for that software component or microservice in its external configuration file or system.

5. **Build, release, run**: *Strictly separate build and run stages.* This principle mandates a separation across the software build, release, and run stages. Using the aforementioned principles, you can have a CI/CD pipeline per microservice. In that pipeline, you configure the different release stages for the microservice. Typically, the artifact of the microservice pipeline generated at the end is tagged using the agreed release version format (for example, release-major-version.release-minor-version.patch-number, such as v1.0.0). The artifact could be a Docker image of the microservice tagged like this.

6. **Processes**: *Execute the app as one or more stateless processes.* This principle mandates that each microservice should run in its own process isolated from other running services. We talked earlier in this book about side-car containers, where we learned that the microservice core business logic runs in its own container, while other cross-cutting aspects, such as logging, auditing, and security, can be run in another container. Both containers run inside a single Pod if you use Kubernetes as the container orchestration engine. Or, as another example, returning to the backing service, this is run in a different process (or typically on a completely different node) than the process that runs the microservice itself. Later in this chapter, we will talk about the **service mesh** concept, which aligns with this principle.

    The end goal of this principle is to make sure your microservice is 100% stateless to get the benefits of the cloud elasticity model explained earlier.

7. **Port binding**: *Export services via port binding.* This principle mandates that each microservice should expose its interfaces (APIs) on its own port.

8. **Concurrency**: *Scale out via the process model.* This principle mandates that each microservice should be scaled out smoothly. If you've already followed the previous principles, this principle will eventually be achieved. In other words, if you develop a 100% stateless microservice that is fully decoupled from other microservices and backing services, runs in its own process, and is self-contained, then scaling out (by adding more nodes/servers or containers to host that microservice) is straightforward. This principle emphasizes the fact that cloud-native apps should support scaling out (horizontally scaling), not scaling up (vertically scaling).

9. **Disposability**: *Maximize robustness with fast startup and graceful shutdown.* This principle mandates that service/microservice instances should be disposable, meaning the service instance should start up and shut down fast. Using container technologies helps achieve this principle.

10. **Dev/prod parity**: *Keep development, staging, and production as similar as possible.* This principle mandates keeping environments used across the application life cycle as similar as possible. In traditional software development practices, we used to hear, "It was working on my machine or in the dev/integration server – I don't know why it's not working now in the production server!" Cloud-native apps should be infrastructure-agnostic and self-contained, so they should run locally and in the dev/staging environments just the same as in the production environment, or any other environment. Again, container technology helps achieve this principle. If you have your cloud-native app as a container image (let's use Docker as the container type here), then you can run that cloud-native app wherever Docker Engine is installed, with the same experience every time.

11. **Logging**: *Treat logs as event streams.* We explained this principle earlier as an example of the differences between traditional and cloud-native software.

12. **Admin processes**: *Run admin/management tasks as one-off processes.* This principle mandates running administration tasks, such as data cleanup and computing analytics, as one-off processes that are separated from the core cloud-native app code.

Those are the famous Twelve-Factor App principles. There are some other extra factors or architecture principles that are used besides the cloud-native Twelve-Factor principles when designing and building modern apps, such as the *API-first architecture principle*.

This mandates that everything should be exposed as an API, and is a great architecture principle for building modern apps as it opens the door for so many existing frontend consumers and clients, such as web apps, mobile apps, and generally any HTTP-enabled device or client that can interact with your business HTTP or REST APIs.

By going API-first, you cover not only the existing and well-known frontend consumers and clients, but also future, unknown, and upcoming types.

Another extra factor is the telemetry design pattern. Note that this is not the IoT telemetry design pattern we discussed in *Chapter 1, Introduction to The IoT – The Big Picture*, but it does have the same concept mandated by that telemetry design pattern – that you must **observe** and **monitor** your software or microservice using different monitoring tools and check-ups.

There are yet more extra factors, and more patterns keep emerging every day, so some people prefer to call it the 12+ factors.

Public cloud providers, such as AWS and Microsoft, provide what is known as a Well-Architected Framework, which comes with a set of industry-standard best practices that help in improving the quality of cloud-native workloads deployed in those public clouds.

The AWS and Microsoft Well-Architected Frameworks share similar concepts for a well-architected workload, namely, cost management, operational excellence, performance efficiency, reliability, and security.

We recommend you apply a Well-Architected Framework while designing, building, and running your cloud-native apps, especially if you run your workload in those public clouds. Well-Architected Frameworks can be used with any other cloud, though, including private clouds, as these frameworks usually start with generic concepts and best practices, before then explaining what Amazon (in the case of AWS) or Microsoft (in the case of Azure) provides in terms of tools and technologies to tackle those concepts and make sure your workload is well architected.

With cloud-native now defined and a quick explanation of the Twelve-Factor App methodology under our belts, it is clear you need to follow these principles in both the design and implementation stages to build a modern cloud-native software or app. We saw some technologies and architecture styles that can help you achieve this cloud-native vision, including microservices, API-first, containers, backing services, automation, and CI/CD pipelines.

We covered containers earlier in the book, so in this chapter, we will move on next to the microservice architecture.

Microservice architecture

For the last few years, up to the time of writing this book, everyone has been talking about microservices, even though they've been around since 2011. It is clearly a hot topic in the software industry community.

There are so many definitions of the microservice architecture, and we will come to one of them eventually toward the end of this section. However, we would first like to share a story of how we ended up with a microservice-like architecture in one of the large-scale e-commerce platforms that we built around the same time the term microservice started to become more prevalent in the software industry community.

The e-commerce platform was built using a framework based on standard Java EE technologies.

The architecture of that e-commerce platform followed the standard N-tier application architecture, namely, the three-tier architecture. In this architecture, there is a presentation layer (UI), which mainly contains the web pages for the application, such as the profile, product, cart, order, and checkout pages; the business layer, which contains the business and data access logic for the whole application in the form of Java classes or Java Archive (JAR) files; and finally, the data store layer, which is an Oracle database with one or two database schemas.

Everything worked fine with that type of architecture and in fact, the architecture was state of the art at the time. The e-commerce application was also integrated very well with the organization's other services, including the payment gateway, address and identity verification, and delivery/shipment gateway, through a well-defined SOA layer.

In those days, around 2011, moving to a public cloud and using cloud computing's novel elasticity model (that is, scaling out or scaling up easily, autoscaling, and self-healing) wasn't so widely done in large organizations and enterprises. At that time, large organizations used their data centers to run their business software applications, so system scalability was simply done manually or was static. We used to have a proper capacity planning exercise to evaluate how much additional hardware we needed based on the consumer traffic forecasts and the number of different consumer channels (such as traffic from web apps, mobile, and point -of-sale machines) and other business factors. With this knowledge, we made purchase orders for the required hardware a few months before the time we expected the additional traffic so our e-commerce platform would be ready to scale.

The main challenge or problem we faced in that large-scale e-commerce application was scalability, especially on seasonal occasions, such as Boxing Day and Cyber Monday, and when offering pre-orders or launching new products.

Despite our capacity analysis and planning, we still struggled with the scaling of the e-commerce platform to absorb all customer requests. Sometimes we had to throttle the number of requests to avoid the whole application going down due to the overload, but in business terms, those requests meant orders and hence revenue. So, our **risk assessment strategy** was "losing a small percentage of the orders is better than losing all orders." If you ask the people on the commercial side of the business, they don't want such throttling and the loss of orders, but that was the technical challenge we faced, and it lasted only for a few hours or a couple of days. After that, the platform acted normally as the load was reduced, but now we had **idle resources** (namely, the hardware we purchased) so it was a waste of money!

To solve that scalability issue, we did an analysis and found that the majority of customers accessing the e-commerce website during those seasonal events spent most of their time on a few pages such as product pages, where they could browse product details, customer reviews of that product, and other recommendations. Interestingly, most customers abandoned their visit to the e-commerce website at that stage and never proceeded to the checkout pages. We discovered they proceeded to stores on the high street to purchase the product if they liked what they saw online.

Now, this is interesting, as you a set of pages and functionality of your website (the product pages) are under a lot of stress, while other pages (the checkout pages) are not under much stress.

The e-commerce website was packaged as an **Enterprise Application aRchive (EAR)** containing the *presentation*, *business*, and *data access layers* and deployed redundantly on several virtual machines. To scale, we simply added additional virtual machines and deployed the EAR file of the entire e-commerce application functionalities, including both the product and checkout pages. But remember, based on demand, we only needed to scale the product pages, so it wouldn't make sense to keep scaling the checkout pages as well.

They were not suffering as the product pages were and it would just waste resources.

At that point, we needed to think about splitting our whole e-commerce application into two parts or two applications, one for the **product journey** and the other for the **checkout journey**. This way we could scale them independently, solving our problem using fewer resources (we only purchased hardware for product service scalability) more efficiently.

What you just read is one example of what microservice architecture can solve. We only focused on solving scalability issues and didn't go beyond that, but microservice architecture can solve other issues as well. For example, it allows you to develop the microservices separately, meaning you can use different programming languages or paradigms to develop each service. In our case, we were focused on the deployment and scalability aspects alone, hence we used the same technology stack for both the product and checkout functionalities.

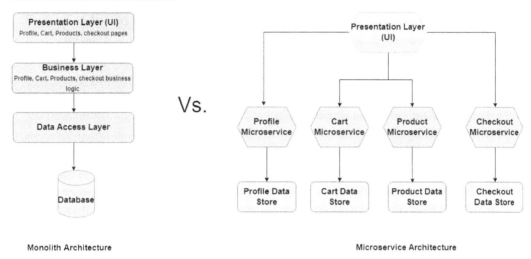

Figure 8.1 – Monolith versus microservice architecture

Now, we can say that microservice architecture is an architecture style (not a technology) that breaks an application down into a smaller set of independent services (called microservices) that interact with each other to deliver business user journeys and/or end user requirements.

We can summarize the benefits of microservice architecture in the following points:

- This type of architecture tackles the problem of application complexity as it enforces more modularity, usually leading to small services that are easy to manage and quickly understood by application developers. Microservices are easy and fast to develop and maintain due to their size.

- Because the service is small, re-writing the whole service again becomes feasible and easy. So, if you make the wrong decision or direction (such as choosing the wrong technologies or inappropriate designs) when designing or building the microservice, you can revert it quickly and build a new microservice from scratch. Compare this to monolithic applications where rewriting means wasting lots of effort, time, and money.

- The service can now be developed independently, so you have the freedom to choose the right technology stack, tools, and frameworks for your microservice. This is a great benefit as you might have some microservice that requires more memory management, in which case you could build it with C or C++, while other microservices could be built with Java, Python, Go, or .NET, among others.

  We said this is a great benefit; however, in real life, we usually recommend sticking with one or two technology stacks for building microservices. Why? To keep a degree of control, as if you leave it open for different teams to choose the technology stacks, then you might end up with every team choosing different technologies. This will be a problem in the long run in terms of finding developers or engineers with such varied skills to maintain and develop the microservices. So, from an organization governance point of view, it's worth mandating specific frameworks and programming languages to be used across the organization and the different teams.

- *Fault isolation*: In monolithic applications, if just one functionality causes a problem, such as a memory leak, then the whole application will suffer as all functionalities share the memory and other compute resources. In a microservice architecture, services are isolated and each runs on its own resources, so if any microservice is impacted or has a problem, it will not impact other microservices in the application.

- *Data isolation*: As with the benefit of fault isolation, the microservice architecture also recommends one data store per microservice. As a result, the data used by the microservice is isolated and protected from the data used by other microservices. Thus, the microservice architecture provides isolation on the compute part (see the preceding fault isolation point) and the data storage part, which is a great benefit for any modern app architecture.

- *Services can be deployed independently*: This is a very important benefit, as you can make updates to any microservice without impacting other microservices of the application. In real life, you might have an application containing a set of microservices running different software versions of the same microservice, such as AccountService V1, AccountService V2, ProductService V1, and ProductService V2.

- *Services can be scaled independently*: We explained this benefit when we talked about our e-commerce website challenge.

- Microservice architecture is more agile than other application architectures, such as the N-tier architecture. This agility helps when launching products and apps quickly to the market.

- The microservice architecture fits very well with modern CI/CD pipelines used to deliver more frequent releases in a more automated way.

The microservice architecture is not a silver bullet, nor the only answer to the question, which application architecture we should use? Yes, it is a hot topic in the industry, but our recommendation is to assess first whether microservice architecture will solve your problem or not. In some cases, the microservice architecture might be a killing point and you should avoid it. Ultimately, it is like any other architecture: it has its pros and cons and you, as the solution architect, are the only one who can make the decision about whether to implement it or not, as you have the required knowledge about your context and the inputs at hand.

The following are some challenges associated with the microservice architecture:

- Microservice architecture has the same common challenges that come with any distributed systems architecture, such as the following:

  - **Log tracing**: Tracing logs in applications composed of microservices is complex, as the microservices might be run on different servers and networks in different time zones, typically making log correlation difficult.

  - **Network latency**: With microservices distributed across different networks, there will be latency challenges as communication is done over the wire.

  - **Network unreliability:** In distributed systems, there is no 100% guarantee of network reliability, so at any point in time, you should expect a disconnection or failure to happen to the underlying network. Microservice-based applications are typically distributed across different nodes and networks so communication between microservices or even between microservices and other downstream or upstream systems might be broken because of that network unreliability issue.

  - **Auditing and security**: Auditing, or knowing who did what, in the microservice-based architecture is a bit challenging as typically, such microservice-based applications have hundreds of distributed services/microservices, or even more. With each service, tracking what is allowed/not allowed for different users and systems is not an easy task. On the security side, by having so many microservices and without proper control, you might end up with different security mechanisms used by each microservice; for example, some microservices expose their APIs without any security control, others with basic authentication, and others with OAuth 2.0.

- **Overall application performance/microservice chattiness**: You might need to call multiple microservices to deliver the functionality required for just one web page. This will impact the performance of the application if not implemented correctly.

- **End-to-End (E2E) testing** is a challenge with an application built using the microservice architecture as the services are developed and deployed independently.

- If you have an application that requires transaction management across different microservices, handling that transaction in a microservice architecture is challenging as the microservice architecture recommends a "one database per microservice" pattern.

To clarify this further, a transaction, such as a money transfer from one account to another account, is typically made up of multiple operations (for example, in the case of a money withdrawal operation from the first account and a money credit operation to the second account, all transaction operations must be done or completed successfully; otherwise, in the event of an error in any of the transaction operations, all transaction operations, both those that are successful and have been completed and the failed ones, must be rolled back for the sake of data integrity. In other words, a transaction must be **Atomic, Consistent, Isolated, and Durable** (**ACID**).

Handling transactions in a single service/microservice is easy (such as the example we mentioned earlier about money transfers between different accounts that are typically handled by a single microservice), as you can make sure the transaction is ACID at either the application level or database level if it is a single database.

Handling a transaction across different services/microservices, such as placing an order transaction in an e-commerce microservice-based application, is complex since the transaction operations, in that case, are distributed in four different microservices. For example, creating an order operation is typically part of the order **management microservice**, processing a payment operation is part of the payment **management microservice**, updating an inventory operation is part of the inventory **management microservice**, and delivering an order operation is part of the delivery **management microservice**. Do note that we only choose a very simple order processing flow/ transaction; there are more complex order processing flows that might have more than four microservices participating in their order processing transaction.

To ensure a successful order processing transaction, all four microservices must complete their respective operations successfully. If any of the microservices fail to complete their operation, all of the completed and successful preceding operations of other microservices must roll back to ensure data integrity.

- The frontend UI for the microservice-based application might be considered a bottleneck as there is one common UI across all microservices. The different microservice teams can do the business logic part (such as APIs) on their own without having any dependencies on other teams, but when it comes to the UI part, they have to wait for the team responsible for the UI layer to update their changes.

- Microservice architecture also poses some organizational (non-technical) challenges:

  - **Significant operations overhead**: Instead of supporting one full-fledged application with one technology stack, now you have many small applications with different technology stacks that require support and monitoring.

  - **Skills required**: Applications built on the microservice architecture require a different set of skills that might be rare among developers. Ideally, people with full stack experience and DevOps skills are the best fit for teams building microservice-based applications.

Those are some of the challenges in a microservice architecture; however, there are also many solutions to overcome those challenges. For example, for the transactional control required in microservice-based applications, you can implement the Saga pattern. For the single-UI challenge, there's what is called the micro-frontend. To deal with chattiness, you can use an API gateway for aggregation and build the required backend APIs to meet the required **User Experience** (**UX**) (such as mobile or web). To deal with network unreliability, you can implement the retry mechanism or circuit breaker software pattern.

Explaining each challenge and how to overcome it is beyond the scope of this book, but our view is broadly that while there are challenges associated with the microservice architecture, there are also solutions. In other words, the journey toward a full microservice architecture will not be easy, but you will reach your destination in the end.

Now, design-wise, each individual microservice should do the following:

- **Have a single business or technical focus and responsibility**: In other words, it should represent a business function or domain. For example, an accounts microservice is responsible for anything related to the accounts domain or entity in the application. Likewise, the product microservice is responsible for anything related to the product domain or entity in the application, and so on.

- **Follow the smart endpoints and dumb pipes design principle**: In other words, the internals of the microservice are a black box; the most important thing is the microservice's interfaces or APIs. Communication with a microservice's interfacesAMQP is done through standard communication protocols, such as HTTP/HTTPS, gRPC, WebSocket, or AMQP.

- **Be 100% stateless**: As explained earlier in the cloud-native section, the more you keep your microservices and business logic stateless, the more easily you can scale that microservice and become more resilient.

- **Be independently changeable and deployable**: Remember our earlier example of the e-commerce website? The product microservices were completely independent of the checkout microservice, so scaling a specific microservice could be done easily, and changes in one microservice could be made without impacting other microservices. By contrast, in monolithic applications, small changes mean a full build and regression test on the whole system.

- Be loosely coupled with other microservices.

- Embrace failure and be 100% resilient.

- **Be monitored and tracked**: The microservice should be integrated with central monitoring and logging platforms.

- Be autonomous to integrate easily with CI/CD pipelines.

- **Be cloud-native**: we explained what this means earlier.

- **Have clear, well-defined, and well-documented interfaces**: If the microservice exposes its interfaces as REST APIs, then you could use Swagger, for example. The point at the end is to have well-documented interfaces to simplify them for other teams who consume those microservice interfaces.

- **Be aligned with the REST API model in terms of Create, Read, Update, and Delete (CRUD) operations**: For example, HTTP GET is mapped to reading operations, and HTTP POST to create operations.

Building or implementing microservices is not a big deal thanks to the many microservice frameworks out there that help you build a microservice in just a few minutes. With these frameworks, you can focus on your business logic while other cross-cutting concerns or technical enablers, including circuit breakers, event sourcing, service self-registration, log emitters, and embedded web servers, will be taken care of, and even come out of the box with some microservice frameworks.

These frameworks include Spring Boot and Spring Cloud (Java), JHipster, Eclipse Vert.X (Java), GoMicro (Go), Molecular (Node.js), and Lightbend Lagom (Java or Scala).

The challenge, as we have found in our practical experience, is not building the microservice on its own, as building microservices is super easy, but typically in building the complete E2E architecture of such modern and cloud-native applications. Such applications not only have microservices in their architecture but also have components such as gateways (API gateways and proxies), backing services, load balancers, identity and access management, monitoring and alerting, and container orchestration systems such as Kubernetes.

With that, you should now understand at a high level what microservice architecture is, its benefits, its challenges, and what factors should be considered when designing microservice-based applications. In the next section, we will cover aspects related to microservice architecture that are often contentious topics, starting with microservice communication.

Microservice communication

You have built the microservices using one of the frameworks we mentioned in the previous section, and now the question is how the external frontend clients, such as web and mobile apps, will interact or communicate with those microservices. Also, how will the microservices communicate with each other internally (if needed)?

For external frontend clients, you have the following options:

- **Direct communication with the microservices**
- **Communicating through API gateways**

The option of direct communication with the microservices is not recommended at all due to the following:

- It is not secure – to ensure security, you must not expose internal microservices to external clients.

- External frontend clients' code will be more complex as they must manage interactions with all microservices' endpoints, where each microservice might have different API interfaces (one might expose REST endpoints, others expose SOAP or gRPC, and so on) or use different authentication methods.

- It couples the frontend clients tightly with the microservices, reducing your ability to make changes and release new features in the future.

The API gateway option is the recommended and widely used option for communication in the microservice-based application architecture.

There are many benefits of having an API gateway in front of the internal microservices:

- Since it is a gateway, it offers more security and resilience features to your applications. Most API gateways offer the following features out of the box, usually through policies applied to incoming and outgoing traffic as it moves through the API gateway:

  - **Throttling**: You can throttle a specific client, or have base throttling on the number of calls per minute, among other parameters.

  - **Payload size limit**: You can protect your backend service by restricting the size of the accepted payload in the exposed APIs.

  - **Spike arrest**: Similar to throttling, you can protect your backend services against severe traffic spikes and denial -of -service attacks.

  - **Rate-limiting**: You can put a hard limit on your exposed APIs to protect your backend service so the client in this case will follow that limit. For example, if you put a hard limit of 90 Transactions per Second (TPS) for the APIs, the API callers (the clients) know that if they exceed that number of TPS, they will get a rate limit error, while in the case of throttling, the server might allow the throttled request/ transaction (for example, transaction number 91) if some space/capacity becomes available on the server side.

  - **Content-based security**: An API gateway works in layer 7, or the application layer, so you can easily read and intercept the contents from the HTTP headers, parameters, URI, and body and act accordingly, such as reject or block the request, or count or allow the request.

  - **Mutual SSL or two-way SSL**: Authenticate API clients and consumers using X.509 certificates.

- **Access control**: Most API gateways on the market offer access control for accessing the APIs, including OAuth 2.0 (the most used option nowadays), SAML, basic authentication (username/password), API keys, and custom authentication and authorization.

- **Monitoring and tracking**: If, in your architecture, your API gateway is the front door for accessing any microservice, then you can solve the traceability/auditing challenge mentioned earlier. The API gateway generates a unique transaction ID that's passed over to the called microservice, which passes it over to the next downstream system that's called, and so on. This provides visibility in terms of how the request traverses from the frontend gateway through all the application microservices and downstream systems called to fulfill client requests.

- **Analytics and KPIs**: You can collect different KPIs and metrics from the API gateway for analysis. This includes performance KPIs, such as how many TPS or transactions per minute your application can support, the most-called APIs, the least responsive APIs, the top faulty APIs (APIs with the most errors and business faults), and API usage.

- **Backend for frontend pattern**: This is an important benefit of using API gateways in microservice-based applications. By using this pattern, you build what are known as experience APIs. Let's give an example to help you understand. Imagine you have some UI/UX differences in your application depending on whether it's accessed from web browsers, mobile browsers, or mobile-native apps. For example, the screen is big enough in a web browser that you can display all details about a product on the product page, while in the mobile experience, you need to show just a few details in a more condensed format to fit the smaller mobile or tablet screen sizes.

  By using an API gateway, you can expose one API (the `Product_Web` API) to be used only for the web browser clients and another API (let's call it the `Product_Mobile` API) for the mobile journey. The `Product_Web` API might return XML, while the `Product_Mobile` API returns JSON. The point is, the backend APIs of the microservice are the same in both instances; you just build wrapper or proxy APIs on top of the backend API to fit the UI/UX you need, without changing the backend APIs.

  This pattern overcomes the challenge of microservice chattiness, as the API gateway receives one request from the frontend, internally calls the different microservice APIs, aggregates the results, and returns this to the API caller.

  You can have a dedicated API gateway per client (such as a mobile API gateway, web API gateway, and desktop API gateway), or you can have one single API gateway that hosts different experience APIs relevant to each client or channel.

- **Supporting different payload and protocol transformations out of the box**: For example, your microservice might expose its backend APIs in SOAP, but you have adopted a modern frontend client that requires APIs to be in REST format. Instead of creating a new API from scratch to meet that requirement, most API gateways on the market now support SOAP-to-REST transformation out of the box. So, you can keep your microservices with the old SOAP APIs and just expose the REST APIs in the API gateway for the modern external clients.

As another example, your microservices API could support REST but use XML for its payloads, while most modern frontend clients require JSON payloads. Here, an API gateway can be used to provide an out-of-the-box transformation from XML to JSON and vice versa.

Before we mention some of the API gateway vendors, you should first understand the difference between an API gateway and an API management suite. An API gateway is one component of the management suite, which typically contains the following:

- **API gateway**: This is the main core component of the API management suite and handles the incoming and outgoing API traffic.

- **API analytics**: This provides an out-of-the-box API analytics solution.

- **API developer portal**: This usually offers the following features for API developers and consumers:

  - API online interactive documentation (interactive in the sense that you can explore the API documentation in an online UI, and can fire some API requests and see the response from the online portal).

  - Discovering, testing, and subscribing to published APIs.

  - API developer forums.

  - Downloading the sample code and SDKs for fast integration with APIs.

  - Managing client keys and secrets used for accessing APIs.

  - API publisher or admin portal: This portal is used by API providers to manage their exposed APIs. When APIs are published using this admin portal, they become discoverable in the API developer portal to developers with the right access.

  - API monetization or billing: Some companies sell their services in the form of APIs, so they treat their APIs as products that have the complete life cycle of a sellable product. API monetization and billing components help companies sell and manage their API consumers' subscriptions (such as by offering subscription levels of Developer, Basic, Premium, and Gold, where each level offers different features).

There are many API management vendors on the market offering open source, cloud-based, or commercial API management solutions, including Apigee (Google), MuleSoft (Salesforce), WSO2 (open source), Kong (open source), Software AG, TIBCO Mashery, Amazon API Gateway (AWS cloud), Microsoft Azure API Management (Azure cloud), IBM API Connect, and CA Technologies.

You might only need an API gateway component for your architecture, without the full management suite. If this is the case, there are many open source, lightweight API gateways, such as Spring Cloud Gateway, NGINX, Zuul, and Linkerd.

API gateways are an important component in your microservice-based application architecture, and you might ask, *Which one should I use? A complete API management suite or just a lightweight API gateway?*

To answer that question, you need to assess whether you have requirements related to API monetization, API developer portal experiences, and API analytics. If you do, then you need a proper API management suite, not just an API gateway. But if not, then just a simple, lightweight API gateway should be enough for your microservice-based application architecture.

By now, it should be clear that the right approach for frontend clients to access microservices is to use an API gateway to control that access and communication, but what about microservice-to-microservice communications?

There are two types of communication: **synchronous**, where the client sends the request and waits for a response to that request, and **asynchronous**, where the client sends the request and doesn't wait to get a response back. With the latter, the response to the request can come later through a callback mechanism, or the caller can keep checking to find out whether the response is ready or not.

If your microservices communicate asynchronously, then you have an easy life – you can simply use some broker, such as Kafka or RabbitMQ, between those microservices. Each microservice sends an event to a specific topic and other microservices can subscribe to those topics to receive relevant messages. In other words, this is the very frequently used publish/subscribe mechanism, also seen in event-driven architecture.

Asynchronous communication is the recommended communication pattern in microservice architecture, but we all know that real life is often not that simple, and the microservice might need to call another microservice or another downstream system synchronously, not asynchronously.

If microservices communicate through the synchronous pattern, then we have the following options:

- **Direct communication**: This is not recommended as it couples microservices to each other, which is not good design practice.

- **Using an internal API gateway**: This will solve the problems of direct communication between microservices. It is a pattern for external frontend clients accessing internal microservices, so you end up having two types of API gateways in your architecture: one for external integration and one for internal communication.

- **Using a service mesh**: This is a modern, rapidly evolving technology used for internal microservice communication, which we will examine in the next section.

Service mesh

If you look deeper into the anatomy of a microservice component, you will see that at the core of it is the business logic written in a chosen programming language. The business logic of the microservice typically covers the CRUD operations plus some other custom business logic operations.

The microservice business logic is what the microservice developer should focus on and ultimately deliver, but when developing software with microservice-based architecture, there are lots of other cross-cutting concerns that should be covered by the microservice as well, such as logging, security, auditing, resiliency patterns such as circuit breakers, monitoring, configuration, service registration and discovery, communication, and networking services.

Fortunately, there are many microservice frameworks and other open source cloud-native projects available on the market to handle these cross-cutting concerns out of the box. This lets microservice developers focus only on the business logic of the microservice, nothing else.

Now, how will those cross-cutting concerns be implemented in the microservice-based application architecture? Typically, those cross-cutting concerns are implemented in microservice architecture by using the **sidecar design pattern** with one of the cloud-native design patterns.

The idea of the sidecar pattern is simple. This pattern mandates that cross-cutting concerns and any non-core functionality of the microservice should be encapsulated and run in another separate process (that is, not the core microservice process). It is named sidecar as it is conceptually similar to the sidecars attached to motorcycles (see *Figure 8.2*). These sidecars go everywhere the motorcycle goes, and are dependent on and coupled with the motorcycle – in other words, its life cycle shares the main motorcycle's life cycle, being created and retired alongside the main motorcycle:

Figure 8.2 – A physical sidecar attached to a motorcycle (source: https://wordpress.org/openverse/photos/43c8df05-2a09-4c2f-beff-3dd193b322d7)

So, in short, when deployed, your microservice component will have the main business logic process plus the sidecar process that handles the cross-cutting concerns.

The sidecar pattern is often used with containers. In Kubernetes, as explained earlier in the book, you can have one Kubernetes Pod that contains two containers running inside that Pod: one container (or process) runs the main business logic of the microservice, and the other container runs the sidecar handling the cross-cutting concerns. Since both containers are run in the same Pod, they can communicate with each other through localhost (as they are just two different processes sharing the same host).

This is the concept of the service mesh, and it provides lots of architectural flexibility and extensibility, especially with something such as Kubernetes. Consider frameworks such as Fluent Bit and Fluentd, used for extracting logs from servers or containers and sending them to logging platforms such as Elasticsearch – those frameworks also use the concept of the sidecar pattern. This means you can have your microservices deployed as containers in Kubernetes (where each microservice has its own YAML file describing the container deployment specs for that microservice), and you can simply add more containers or co-locate them to the main containers to extract the logs from those main containers. This means you don't touch the main container that hosts the given microservice – rather, log extraction is done in another process. You can do this extension even after the main container is already deployed – awesome, isn't it?

Service meshes use the concept of a proxy service. The proxy is co-located with each microscrvice, as explained previously, so all inbound and outbound traffic for that main microservice must go through that proxy, which will handle all cross-cutting concerns, such as routing traffic and logging.

One interesting feature of traffic routing provided by service mesh systems is that you can split the traffic between internal microservices. For example, the service mesh/sidecar process (proxy) can split the traffic between two different versions of the same microservice. This could be done to test a new version of the microservice, where the service mesh will route, say, 20% of the internal traffic to the new version of the microservice and the remaining 80% to the old version.

There are many service mesh implementations on the market, both open and closed source, such as Istio, Linkerd, Consul Connect, Kuma (from Kong), `Maesh`, `servicecomb-mesher`, OpenShift Service Mesh by Red Hat, and AWS App Mesh.

The most famous and commonly used one is Istio, built on the Envoy proxy. Envoy is an open source edge and service proxy designed for cloud-native applications. Indeed, most of the preceding service mesh services are built on or support the Envoy proxy.

So, you might ask, which option should I then use for microservice -to -microservice communication? The direct communication option, the internal API gateway, or the service mesh?

To answer that question, we recommend using API gateways to handle the external traffic in the direction of the microservices, the service mesh to handle the synchronous communication between internal microservices, and a message queue or a broker such as Apache Kafka to handle the asynchronous communication between internal microservices.

With this, we have covered the topic of microservice communication. In the next section, we will continue looking at other hot topics in the microservice-based architecture. Let's talk about API formats for external and internal APIs.

API formats

There are two types of APIs in the modern cloud-native application architecture, external APIs and internal APIs. Internal APIs are those that are exposed by the microservices, which we sometimes refer to as backend APIs.

We mentioned earlier that microservice architecture follows the smart endpoints, dumb pipes concept, so you will never see a microservice without a set of APIs that covers its functionality.

External APIs are those APIs that you expose for internal and external customer integration. External customers are typically customers who would like to call your exposed APIs to build – by themselves or through their partners – their web, mobile, or whatever UX they want at their side, while internal customers could be your company's internal applications, such as the company website or mobile app.

The question now is, what should external and internal APIs look like in terms of communication protocol, payload, and so on? Is it the same for both internal and external APIs, or are there differences?

Let's take it one by one. When it comes to exposing APIs, typically you use REST, SOAP, or **Google Remote Procedure Call** (**gRPC**) API styles. We will not discuss the pros and cons of each API style here as it is beyond the scope of this book; however, we will give our recommendations and thoughts based on our practical day-to-day experience of architecture and design discussions with different development teams.

We recommend exposing external APIs as REST APIs. Why? First of all, forget about SOAP – it's become an old, legacy API style, and is heavier and more complex than REST APIs, so there's no point in discussing that option. The only time you would expose SOAP APIs externally is when your customers already have external SOAP APIs and don't want to migrate to more modern REST APIs. Otherwise, most modern application architecture exposes external APIs as REST APIs.

The competitor to REST APIs is gRPC APIs, so let's understand a bit about gRPC.

gRPC is the modern version of an old technology concept called **Remote Procedure Call** (**RPC**). There are different technologies that implement that RPC concept, such as Java **Remote Method Invocation** (**RMI**).

The idea of RPC is that you have a software client running locally (that is, in the same process as the main application process, which would happen if you reference the software client code in your main application code). The software client exposes different business methods and operations of a remote service. Now, in your main application code, you just call client methods/APIs locally as if you were calling the remote service APIs. Then, under the hood, that local client or stub will make calls to the remote service that has the real logic of those business operations or methods. Also under the hood, the local client or stub will take care of networking complexities, serialization, and so on.

gRPC is well known across the cloud-native community. It supports the most popular development stacks, including Java, JavaScript, C#, Go, Swift, and Node.js, so you will easily find a gRPC library for your microservice.

But why gRPC?, you might ask. Well, it is lightweight, highly performant, and much faster than REST/JSON APIs. gRPC uses HTTP/2 for its transport protocol, thus inheriting the following benefits of the HTTP/2 protocol:

- HTTP/1.1 is a text-based protocol, while HTTP/2 is a binary protocol, making it much faster and more efficient for data transport.

- HTTP/2 supports multiplexing for sending multiple parallel requests over the same connection, making it faster and more efficient compared to HTTP/1.1, which is limited to one request/response at a time.

- HTTP/2 supports header compression, which helps with limited network bandwidth.

- HTTP/2 supports bidirectional full-duplex communication for sending requests from the client and receiving responses from the server at the same time.

- HTTP/2 supports asynchronous communication between the client and server.

The greatest benefit of gRPC is data serialization, as messages are serialized as binary bytes. This makes it super-efficient compared to the JSON or XML data messages used in REST APIs.

gRPC supports Protocol Buffers, an open source cross-platform data format. In gRPC, you basically define the data structure using what is known as **Interface Definition Language (IDL)**. The API contract in gRPC is implemented as a text-based .proto file in which you describe the methods, inputs, and outputs for each service.

Using that .proto file, Protobuf Compiler (protoc) generates both the local client code and the remote service and the messages between them will be serialized into binary bytes.

Without further details, let's jump to the conclusion here about which style of API communication model we recommend for internal microservice-to-microservice communication. Our recommendation is to use gRPC for internal microservice-to-microservice communication due to its super-low latency, high throughput, network bandwidth savings, and real-time communication if needed (where the server pushes messages to clients in real time without the client polling the server every now and then).

We recommended earlier REST APIs over gRPC for external APIs, as REST is a more widely accepted API style than gRPC. REST APIs can be easily called from browsers, while browsers don't yet fully support the HTTP/2 protocol upon which gRPC is built. So, in our view, the right architecture approach is to use gRPC internally to get the aforementioned benefits, and use REST APIs externally for compatibility with all external (HTTP-based) clients available now and in the future.

We have now covered external and internal microservice communication. In the next section, we will cover another topic frequently discussed during the designing phase of microservices-based application development – that is, the databases in a microservice architecture.

Microservices and databases

Any software application typically has two parts, the business logic that is run by the processor and the storage layer that persists the data needed by the application over the long term. In monolithic applications, the storage layer is quite straightforward, with the application and all its features using a shared database. But how do microservices-based applications deal with the database?

One of the benefits of splitting a monolithic application into a set of microservices is that each microservice can select the most appropriate data store for itself. In other words, in a monolithic application, the single shared database is usually of one specific type, either a relational/SQL database or a NoSQL database. But with microservices, you can have one microservice using a relational database and another one using a non-relational database type, such as graphs, documents, and key-value stores.

This is awesome, as you can choose not only any programming language for your microservice business logic but also any database type that you believe is the best fit for the microservice data model. However, you have to be careful here and utilize some design and architecture governance. Earlier, we mentioned that although there's the freedom to choose any technology stack when building microservices, we recommend mandating one or two technology stacks across all microservices to avoid messy organizational issues. The same recommendation applies to database selection, so let's explain this point further.

We do recommend having one database per microservice, but we don't need a situation where the microservice A team decides to go with a relational database, choosing a MySQL database, and while the microservice B team also chooses a relational database, they opt for PostgreSQL, while the microservice C team chooses MariaDB, and so on. This creates too great an operational overhead and will be costly – even if they are open source databases, in a production-grade environment, you will need support for those open source products, meaning purchasing support licenses for MySQL, PostgreSQL, and MariaDB. The same applies to NoSQL databases if a variety is chosen across the teams.

Ideally, we recommend that in your microservice-based architecture, you have two types of databases: one relational database, choosing, for example, PostgreSQL as the database engine, and one non-relational or NoSQL database with MongoDB.

With that single relational database engine option, you can create what is known as a database schema for each microservice that needs a relational database. Likewise, you create document collections or a Table entity for each microservice requiring a non-relational database. This approach allows you to follow the recommended pattern of one database per microservice while still efficiently handling the operational overhead challenge explained earlier.

The devil is in the details, and we can continue discussing design ideas or questions such as "*Why would each microservice have a dedicated schema? That is still overhead; why not just give each microservice a set of tables inside a shared schema between all microservices that use a relational database?*" These types of questions are valid, and this might even be the right solution in your context, but discussing such topics in detail is beyond the scope of this book. Our goal is to share our practical experience and put you in the driver's seat, taking it forward yourself from that point.

With that quick overview of the new, modern architecture style of microservices and their use in building modern apps out of the way, we come to the end of our examination of the microservice architecture. As an IoT solution architect, if you have to design and build an IoT backend – be it the whole thing or just some components for it that are not provided by serverless or ready-made IoT cloud platforms – then you know what you need to start with: designing and building cloud-native apps. This will entail the usage of technologies and architecture styles we've looked at, including containers, microservices, API gateways, and service meshes.

We've tried to cover at a high level as much as we could about cloud-native design patterns and technologies in this book, but there are still many design patterns we have not covered or discussed, such as *Ambassador*, *Anti-corruption Layer*, *Cache-Aside*, *Choreography*, *CQRS*, *Event Sourcing*, and *Strangler Fig*. We recommend you familiarize yourself with all cloud-native design patterns when seeking to design and build robust, scalable, and highly resilient IoT backend applications and solutions.

In the large-scale and production-grade IoT projects that we have designed and built, we followed the following modern technological blueprint that typically covers four areas:

- **Development and delivery processes**: There are so many software development processes and methodologies, chiefly *Waterfall*, *Agile*, and, more recently, *DevOps*.

  The Waterfall process is an old software development process but is still in use in many software design houses.

  Agile was the next-generation software development and delivery process, built on Waterfall. Many organizations have tried to adopt Agile methodologies and the Agile Manifesto as much as they can.

In our practical experience, the best methodology is to combine both Waterfall and Agile processes in the project's software delivery. This means the architecture, designing, and building of the foundational layer (things such as single sign-on and integration with external third parties) follow the Waterfall process, while the user business journeys built on top of that foundational layer follow the Agile process.

DevOps is the latest modern delivery methodology that combines both software development and operations in one single process. In other words, "*you build it, you run it, and you support it.*"

We can't recommend a specific software development and delivery process for your IoT projects, as requirements differ from one organization to the next, based on specific inputs and context. But if you want to follow the latest and most modern approach, then we recommend the DevOps model, although we know that the journey toward DevOps in many organizations is not easy.

- **Application architecture**: There are many application architectures out there, including *monolith*, *N*-tier, *event-driven*, *microservice*, and *serverless architecture*.

  As an IoT solution architect, you should target the event-driven and/or microservice architectures if relevant to your IoT design solution.

- *Serverless* is a much more modern and advanced type of application architecture, but only shines with hyper-scale public cloud providers, such as AWS, Azure, and Google. Serverless may be difficult to employ in private clouds, plus *the portability of serverless-based applications* is a big challenge (that is, if you go with the AWS serverless offering, then porting from AWS to other serverless providers, such as Azure or Google, will not be an easy task, as all of those cloud providers use their own proprietary technologies and tools in their serverless offerings). Hence, we suggest the microservice architecture for large-scale IoT solutions as it aligns very well with the infrastructure-agnostic principle (plus the other benefits of microservices mentioned in this chapter) – you just deploy your microservices into containers, which can run anywhere where there is a container engine.

- **Deployment**: There are many ways to deploy your applications, from *physical servers* to *virtual machines* and *containers*:

  - *Physical* servers are an antiquated and expensive approach and not recommended at all.

  - *Virtual machines* are still widely in use and in fact are the dominant option on the market.

- *Containers (and the tools and technologies in the container ecosystem)* are a new technology that fits very well with modern cloud-native apps and solutions.

  As an IoT solution architect, you should use containers for your deployment, especially if you choose the microservice architecture for your application architecture. Containers are the best deployment option for microservice-based applications.

- **Infrastructure**: There are many different options, from your organization's *private* or *hosted data centers* to the *cloud computing* model, be it a public or private cloud.

  As an IoT solution architect, you should target cloud computing (public, private, or hybrid) as the infrastructure for your applications.

So, in short, for your IoT backend application project, you should target the usage of *DevOps*, *microservices*, *containers* (including Kubernetes and other container technologies), and the *cloud*.

In the next section, let's examine the complete IoT application reference architecture.

IoT application reference architecture

Let's put everything discussed so far into one diagram, creating the typical reference architecture for IoT backend applications:

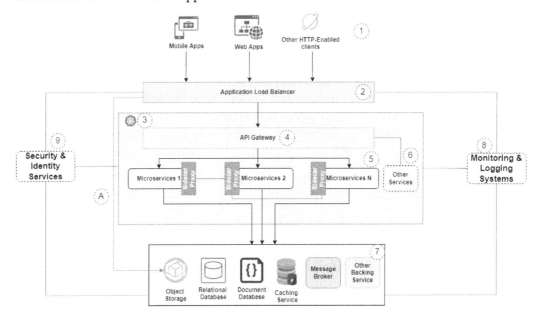

Figure 8.3 – IoT application reference architecture

Let's discuss the main components of this IoT application reference architecture (note that the following point numbers match the numbers in the preceding diagram):

1. **The consumers and frontend clients**: These could be your own frontend clients or external customer frontend clients. Ultimately, both will call the external APIs of your IoT backend application that you have exposed.

   Technology-wise, you could build *native mobile apps* (developed specifically for *iOS*, *Android*, or *Windows*) or you could build mobile apps using *cross-platform mobile development frameworks* such as *PhoneGap, Ionic, React Native, Flutter*, and *Xamarin*.

   Which to choose – native mobile apps or cross-platform development? Each has its pros and cons, and it also depends on business requirements and the budget available. Building a mobile app for each mobile platform (iOS, Android, Windows, and so on) is more expensive than building one app for use across all mobile platforms. In our practical experience, frameworks such as *React Native* are useful to build the experience for mobile and web apps all at once.

   There are many web architecture paradigms *for building web apps,* including traditional web app architecture, such as *Model-View-Controller* (*MVC*) and single-page apps. There are also many web app frameworks to help you build an astonishing web app experience for your IoT solution, including *Angular, React.js, Vue.js, Spring, Django*, and *Flask*.

   The conclusion here is that if you build a robust business logic using microservices, and expose a rich set of REST APIs (following the API-first architecture principle) covering that business logic, then your frontend clients will be the easiest bit in the stack. Ultimately, whether you go with traditional or modern web frameworks, both will consume and use your exposed REST APIs.

2. **Application load balancer**: Traffic on your application can come either from the *public internet* (if you expose your application publicly) or from a *private network* such as your organization's intranet (if your application should only be accessible internally). Traffic from either source should first reach *a load balancer* that acts as a gateway for your application.

   There are typically two types of load balancers. Firstly, there are *network load balancers*, which deal mainly with *layer 4 traffic*. In other words, network load balancers understand only TCP, UDP, and other layer4 protocol traffic. The second type is application load balancers, which deal with layer7 traffic, meaning application protocols such as HTTP and HTTPS.

You need to select the right type of load balancer for your IoT backend application based on your requirements. We will focus on application load balancers here as we are talking about the IoT backend *application* reference architecture.

Application load balancers can help route traffic to the right destination based on the *URI path*, *host*, *header value*, and so on. In other words, these load balancers understand the HTTP protocol, so they can read inputs from the HTTP payload message, use them to evaluate the incoming request, and then decide where they should route the request as configured. For example, you could have one rule configured in the application load balancer to forward any traffic with a URI containing a value such as /api to the API gateway service. This would deal with API traffic, but what about other types of traffic, such as requests for static HTML, JavaScript, CSS, and image files?

To answer that question, point **A** in *Figure 8.3* refers to a direct connection between the application load balancer and the object storage service. The latter could be object storage or a simple web server that hosts the static HTML, JavaScript, and CSS files.

Typically, modern web apps start by rendering static HTML files (such as the login page) that contain the JavaScript files of the relevant framework, such as AngularJS and React. These static files are typically hosted in an object storage layer such as Amazon S3. Such object storage services come with embedded web servers that allow access to the stored objects via URLs, so there is no need for standalone web servers to host static files. Once those files are rendered on the client side (for example, on the browser), then the first part of serving traffic is done – namely, non-API traffic. Then, when the app user clicks a button or a link in that HTML page, and if server-side actions are needed, such as for an API call (for example, https://yourdomain/api/...), then an action will fire. At that moment, the API traffic will hit the API gateway as there's /api in the URI, causing the application load balancer to forward the traffic to the API gateway.

In short, the application load balancer will check the incoming request to see whether it has /api in its URI, and if so, it will forward that request to the API gateway for handling. If not, then the traffic will be directed to your object storage or your web server to handle that request.

Typically, in large-scale and production-grade websites, the domain name of the website has to be resolved to a specific IP. In this context, your project application domain name will be resolved to a load balancer IP, so when users hit `https://yourdomain.com/` then the DNS service will return to the browser the IP (typically more than one IP is used for resiliency purposes) of the application load balancer. The browser subsequently makes the requests to the application load balancer IPs in a round-robin fashion if more than one IP is configured for the load balancer.

3. **Kubernetes**: Your containerized microservice will be managed and deployed into a Kubernetes cluster. We assume your API gateway will also be containerized, hosted, and managed by a Kubernetes cluster to take advantage of the benefits we mentioned earlier in the book.

4. **API gateway:** The API gateway is the most important component in your architecture as it abstracts the microservice backend APIs and exposes the experience APIs for external and internal consumers. We mentioned earlier, when looking at the microservice architecture, that you need an API gateway in front of your microservices.

   You could have one or two logical API gateways, one for external traffic and one for internal traffic. This would be appropriate if you decided to use an API gateway for internal microservice communication or if you had systems that needed to access the internal microservices privately. Alternatively, you could have only one logical API gateway in your architecture with an external network interface configured for external traffic and an internal network interface for internal traffic.

5. **Microservices**: This is where the core business logic is written, creating a bunch of microservices that interact with different backing services and downstream services. These microservices might communicate with each other synchronously using the service mesh concept and the sidecar design pattern, or asynchronously using a message broker such as Kafka.

6. **Other services**: You might need other services to provide functionalities, such as a payment gateway service or GraphQL.

7. **Backing services:** In the cloud-native section, when we discussed the Twelve-Factor App methodology, we mentioned the factor titled Treat backing services as attached resources.

Hence, that layer in the solution will contain all the backing services you need, be they local or external resources – it doesn't matter which, as microservices that follow this factor will access resources using backing service URLs (with access credentials configured externally if required). Microservices can swap from one backing service – say, a relational PostgreSQL database – to another, such as MySQL, without any code change or impact, just by updating the external configuration of that microservice. This relates to another factor of the Twelve-Factor App methodology: *Store config in the environment*. The example here with relational databases could be applied to any other type of backing service as well (such as NoSQL or caching, among others).

8. **Monitoring and logging systems**: Typically, in large-scale and production-grade IoT solutions, you will have a centralized monitoring and logging system that gives you a holistic view of the health status of all IoT solution components.

    Centralized monitoring solutions usually offer a range of monitoring, logging, alerting, and troubleshooting features that help the support and operations teams work with large-scale IoT solutions efficiently.

    In your IoT solution design, you should create a logging and monitoring strategy that covers log generation, log correlation, and the transmission of logs from microservices and other solution components to that centralized monitoring and logging platform. You should also consider how to visualize and get insights from such logs, build custom metrics and alerts, what kind of metrics to monitor (be they infrastructure, service, or application metrics), and so on.

    We will cover more on this topic in *Chapter 9, Production-Grade Operations for IoT Solutions*.

    In short, it is a must to integrate all IoT solution components very well with a centralized monitoring and logging system.

9. **Security and identity services:** Security is a day-zero job, so you must secure your IoT solution by applying the security best practices for IoT devices, edge communication and networking, and IoT backend solutions and applications.

    We will cover more on this topic in *Chapter 9, Production-Grade Operations for IoT Solutions*.

Summary

In this chapter, we have learned what cloud-native apps or solutions are, the difference between cloud-native and cloud-ready solutions, the benefits of building cloud-native solutions, the different pillars of cloud-native apps (including microservices, containers, CI/CD, and automation), and the Twelve-Factor App methodology.

We also examined the microservice architecture, covering what it means and what makes it different from monolithic architecture, along with the benefits and challenges of the microservice architecture. We saw how communication between microservices and with external and internal clients is handled, and looked at API gateways and why they are an important component in the microservice-based application architecture. The concept of service meshes was explained, and we saw how to expose internal, backend, and external APIs in terms of API style (such as REST or gRPC), along with a consideration of the use of databases in microservice-based applications.

Finally, we put all the architecture principles and technologies covered in this chapter into one architecture diagram that we called the typical IoT application reference architecture. We went through the definition of each component used in that reference architecture.

In the next chapter, we will cover the operational side of large-scale and production-grade IoT solutions. We will cover topics related to running and operating large-scale IoT solutions, including monitoring, security, high availability, performance and resiliency, scalability, disaster recovery, backups, and automation.

9
Operational Excellence Pillars for Production-Grade IoT Solutions

Remember that you are on a mission of designing and architecting a production-grade, large-scale IoT solution, not an IoT hobby or fun project. Therefore, you should make sure that your IoT solution is all of the following: *fully secure*; *scalable*; *reliable*; *monitored*; *observed and fully controlled*; *resilient*; *fault-tolerant*; *highly performant*; and *cost-effective*.

IoT solutions are typically complex and contain many systems, components, and layers, so delivering the aforementioned operational excellence aspects is quite challenging.

For example, to make sure your IoT solution is fully secure, you need to start that journey by first securing the IoT devices and microcontrollers, followed by IoT edge devices, IoT gateways, and local networks, before securing communication or the transportation link between the IoT devices and IoT cloud over the wide-area network, and finally securing the IoT backend cloud with all its different components and solutions. Hence, it is not an easy task at all (note, the example is about one aspect only, which is security). On the other hand, in traditional IT systems or solutions, you would most likely spend your time securing (or monitoring or scaling, and so on) only the backend or the server-side part of that IT solution or system (of course, if you offer a mobile app or other thick clients as part of your solution, then you have to secure them as well, but we are talking here about the majority of current IT systems and solutions, which typically have a heavyweight server side and a lightweight client side, like a browser).

In this chapter, we will cover operational excellence pillars, non-functional requirements, and well-architected solution aspects or pillars, while providing some high-level details given that each aspect of those operational excellence pillars may require a dedicated book(s) to explain it in detail. We will explain those aspects in an easy *checklist* form that an IoT solution designer or architect should check during the design and delivery phases of a large-scale, production-grade IoT solution.

In this chapter, we will cover the following topics:

- IoT solution security
- IoT solution monitoring
- IoT solution high availability and resiliency
- IoT solution automation and DevOps

IoT solution security

An IoT security breach is dangerous and scary. In traditional IT systems, the impact and consequences of security breaches are still not considered *life-threatening* impacts. Such impacts (in non-IoT solutions) typically include things such as data privacy breaches or financial impacts, but in the IoT world, the story is different. IoT will share those common security breach impacts of hacked traditional IT systems, plus some other dangerous threats that have a very severe impact on human life (life-threatening). How? Let's talk about an IoT solution in the healthcare *sector*. Think about a patient who has been equipped with lots of IoT devices and sensors that frequently send the patient's biometrics to a healthcare professional to assess the patient's health status and intervene quickly if needed. Now, if those devices have been hacked, what will be the situation if the attacker/hacker sends fake or false biometrics data (note: in that situation, sending fake data is more dangerous than just disconnecting the devices because if the devices were disconnected, then with a proper IoT monitoring system in place, the patient would be contacted immediately, and the device's problem would be sorted out right away). Have you imagined what could happen in that situation?

Not only in the healthcare sector but the *automotive sector* with connected car products – can you imagine what would happen if an attacker gained control of a connected car while you were driving? This is not a fiction movie; we are talking here about real-life scenarios that could happen in our era. Of course, it is great and convenient to control a car remotely, but at the same time, you should think a lot about the scary scenario we have just explained.

In the *utility and energy sectors*, an attacker could cut off the electricity to a whole city or even a whole country if they gained control of the country's or city's electricity grid. We could continue with the impact of IoT security breach examples in each sector or business domain, but the idea and the message are the same in the end: *IoT security breaches are very dangerous.*

To continue, have you heard of the *Mirai Botnet Attack*? It was a very high-volume **distributed denial-of-service** (**DDoS**) attack, known as the largest of its kind in history. It was the largest DDoS in history as it came from IoT devices, which are typically massive in number (thousands or millions) and unsecured due to limitations in computing resources.

Let's explain further why this attack was classified as the largest DDoS attack. In other words, let's study the relationship between IoT and severe and large-scale DDoS attacks?

The idea of a DDoS attack, in general, is to flood or stress the target systems with legitimate traffic (illegitimate traffic would be easily dropped or blocked by firewalls). For example, if an attacker targets something such as an e-commerce website and they need to bring that site down, then the attacker needs a way to *stress*, *overload*, or *flood* the website with *legitimate* traffic. Then, that website will not be able to survive as it has its own infrastructure capacity. Legitimate traffic is normal traffic, such as browsing products and changing passwords, and it comes from IPs with a good reputation and well-known public IPs, so firewalls will accept traffic from such sources unless they are configured to do something else.

The attacker's main task in DDoS attacks is to find or form a *botnet* or a *network of devices* that have a legitimate public IP (not IPs with a bad reputation that are well known to most firewalls and security controls). The attacker then hacks those devices and injects and spreads the malware that the attacker will use later to initiate DDoS attacks on the target (victim) systems.

In a non-IoT botnet or network, the number of devices used in an attack is certainly less than the number of devices used in the case of an IoT botnet as the number of IoT devices is usually massive compared to traditional devices/servers. In DDoS attacks, the rule is as follows: the greater the number of nodes that participate in the attack, the more severe and painful the attack and its impact will be. So, if your IoT devices are unsecured and can easily be hacked and used for DDoS attacks, then it will be the worst nightmare you could think of. In the case of the Mirai attack, it succeeded in bringing down the servers and systems of a company called *Dyn*. This company controls much of the internet's **Domain Name System** (**DNS**) infrastructure, meaning that when it was attacked, large sites such as *Netflix, Twitter, the Guardian, CNN*, and many other sites that use *Dyn DNS* services were impacted and eventually disconnected from the internet, even though those sites might have been up and running – the issue was that they were not accessible to end users because of the DNS issue.

Some large-scale websites can handle, to some extent, DDoS attacks that originate from non-IoT devices by blocking or throttling such DDoS traffic. Even if they cannot block or throttle such traffic, their infrastructure might be designed to dynamically stretch (that is, autoscaling) to absorb such additional DDoS traffic. Of course, this would be costly for some time (due to additional resources being added), but in the end, it would save those websites from going down completely until the security team handles such attacks. While, in the case of DDoS coming from IoT devices, such large-scale websites might not be able to cope with such attacks even if their infrastructure dynamically stretches with autoscaling and different elasticity mechanisms, at some point, there will be no real compute resources available from the cloud underneath to be provisioned dynamically to absorb such traffic.

So far, we have explained the case when IoT devices are hacked and used to participate in a DDoS attack on a target victim. IoT devices could also be victims or targets for DDoS attacks and the impact of such an attack is also dangerous as it might have a physical impact on consumers or users of those IoT devices. Always remember the strong relationship between IoT and the physical world (that is, sensing and actuating). If a website went down for some time, that would not harm the user physically; the user would just not be able to access the website. But if the IoT devices that the user uses serve a medical purpose or any other physical purpose, that could have a physical impact on the user.

In general, the goal of a DDoS attack on IoT devices (the victims) is to keep those IoT devices on (without sleeping) all the time. With that, the batteries of those devices will be fully drained, become flat, and eventually, those devices will be shut down and disconnected.

Another severe impact of a security breach in IoT solutions is a **data privacy breach**. In traditional IT systems, personal data leaked could be data such as a user's credit card information or patient record details. IoT adds additional sensitive personal data at risk of data privacy breaches, such as health biometrics data that is collected from fitness sensors/trackers or medical sensors/trackers; for example, heart rates, sleep trackers, and other personal data where you can't imagine how such data may prove useful to attackers (they could sell such personal data to third parties or they could even use it to threaten the consumer of such IoT devices, and so on).

Here is one example of a smart city solution, specifically a smart waste bin module. Can you imagine that the data collected about a customer's waste bin fullness level could be important data for attackers? Attackers can derive important information from such data. For example, if customers have lots of waste, that could be a sign of how rich those customers are (they have money to spend on food and other stuff, hence they generate lots of waste). The conclusion is that data privacy is a very important aspect to consider in order to have a fully secure IoT solution.

Before we go through our IoT security best practice checklist that IoT solution architects and designers should use and evaluate during the design and implementation phases of an IoT project, let's first talk about some security challenges related to IoT specifically:

- IoT devices are out in the field (or remote). In other words, IoT devices are directly accessed by end users (attackers are a subset of those end users). You might ask, where is the security concern here (that is, how can IoT devices be accessed by end users)? To answer that question, you need to understand the two different philosophies or approaches commonly used in the cybersecurity domain. The first, oldest, and least recommended approach is **security by obscurity**, which means systems or software components will be secured by hiding their implementation. In other words, products, systems, or software components are like a black box. Such systems provide some functionalities to their users, but the users don't know how they have been designed or implemented. The other, recommended approach is **secure or security by design**. That approach promotes the design of systems securely from scratch by following state-of-the-art industrial security best practices during the design and implementation phases.

- IoT devices typically have limited computing resources that make securing them a bit challenging. Security controls such as data encryption/decryption consume or require more computing resources in order to work.

Now, let's move to the list of the security best practices that you, as an IoT solution architect, should use in the design and implementation phases of your IoT solution project. The following list is not exhaustive, but it includes the common practices we follow in designing and building large-scale and production-grade IoT solutions:

- First and foremost, follow the secure by design approach. You should consider applying security principles and best practices in all your IoT solution architecture layers during the design and implementation phases.

- Use and apply a security risk assessment using a well-known framework such as MITRE ATT&CK to collect and evaluate the different security threats that different IoT solution layers could have. The next step typically involves developing a mitigation plan of how to avoid such security threats and secure your IoT solution's layers.

- For securing IoT devices:

    i. Make sure the IoT devices are procured or managed through well-known and highly reputable device manufacturers or **original equipment manufacturers (OEMs)**. In other words, make sure the security and the industrial best practices have been applied across the whole device supply chain. With well-known OEMs, you will get certified secure IoT devices, while cheap devices or microcontrollers that come from unknown third parties are usually not secure, since applying security principles and best practices during device manufacturing and across the whole device supply chain is not a cheap process.

    ii. Make sure to use a real-time operating system that is secure or provides out-of-the-box security controls that can be used to protect the software or the firmware of the IoT devices.

    iii. Never store security credentials or credentials used to connect to the IoT cloud in a plain format inside the IoT device firmware/OS.

    iv. Make sure the IoT devices have a secure booting feature enabled. The IoT devices should only boot with the company-verified and tested firmware or software, not any other firmware or software.

    v. Make sure the IoT device has a unique identity, something such as certificate X.509, and that this unique identity is not shared with other IoT devices through any means.

    vi. Make sure you have a mechanism to control the unique identity of the IoT devices. If the unique identity becomes vulnerable or leaked, you need to be able to revoke, rotate, and manage the unique identities of the IoT devices. A mechanism such as over-the-air updates, which we explained earlier, in *Chapter 6, Understanding IoT Device Management*, is typically used in managing IoT device credentials remotely.

    vii. Make sure you have a mechanism to update the firmware or the software of the IoT devices remotely to patch any security vulnerabilities that could be discovered during the IoT device life cycle (from provisioning to retirement). Over-the-air updates are a typical feature of an IoT device management solution that are used for that purpose.

viii. Don't open ports or install services such as SSH on IoT devices. In other words, reduce the surface area for attacks as far as you can. Why would you open the SSH port on an IoT device? We explained earlier: outbound connectivity from the device to the internet is the preferred method, so if you want to SSH to an IoT device, the IoT device itself can open a WebSocket channel with the terminal that needs access to the IoT device through the SSH protocol. We explained that concept earlier, in *Chapter 6, Understanding IoT Device Management*, when we talked about AWS IoT secure tunneling.

ix. Don't store the customer IoT data that is collected from the IoT sensors in the IoT device endpoint or IoT Edge device's local storage. Always aggregate the important, necessary IoT data and send it to the IoT cloud (that is, the IoT backend). With that step, you protect, to some extent, the customer's data privacy in case those IoT devices get hacked, stolen, and so on. Storing IoT data locally in IoT devices/at the edge is good for handling offline scenarios, but that should be for a short period of time until connectivity is restored.

x. Make sure to apply least privilege permissions. IoT devices usually connect to a cloud or Edge MQTT message broker, so don't give IoT devices wider permissions so they can publish or subscribe to any topic; rather, IoT devices should only have the appropriate permissions required to do their job.

xi. The most recommended method for storing IoT security credentials in IoT devices is to use a hardware secure element that is solderable in the microcontroller board/chip (no one can sniff electrical signals in that case). These are also known as **Trusted Platform Modules** (**TPMs**). By using a hardware security element for storing credentials, you protect the device as, even if it is stolen or an attacker physically accesses it, it will self-destruct once its hardware is touched.

xii. Make sure the IoT device is well-monitored and that you have full visibility of what is going on inside the IoT device. We already covered that in *Chapter 6, Understanding IoT Device Management*, and we explained how to use the logs sent by IoT devices to predict or check any vulnerabilities or anomalies the devices might have. We will cover IoT monitoring in another section in this chapter as well.

xiii. You can use the ready-made AWS IoT device defender solution, which provides full protection for the IoT devices of your IoT solution.

xiv. Make sure to use up-to-date cryptographic libraries and standards, including PKCS #11 and TLS 1.2.

- For securing the IoT local network and IoT edge:

  i. Use a local network connectivity option that is secure. Typically, and as explained in *Chapter 2, The "I" in IoT – IoT Connectivity*, IoT devices will communicate with each other and with the IoT gateway using some sort of short-range wireless network such as ZigBee or Bluetooth Low Energy. Then, the IoT gateway will communicate with the IoT backend over a wide area network (the internet), so you need to make sure the communication link between IoT devices and the IoT gateway is secure to avoid man-in-the-middle types of attacks (someone intercepts the radio signals and gains access to the secure information). Most short-range network options are secure. If the wire-connectivity option (that is, IoT devices connected to an IoT gateway using wires) is used for that IoT local network, then it is more secure, by default, than a wireless option.

  ii. You should make sure the IoT gateway device is fully secure as the device is exposed to the wide -area network (and to the internal or local IoT network as well). You should apply to the IoT gateway device the same security best practices we recommended earlier for IoT endpoint devices. You also need to make sure the IoT gateway device is using a secure communication stack (that is, the data in transit is encrypted, the IoT gateway device validates the IoT cloud server certificate to make sure it communicates with the right IoT cloud backend, and so on).

  iii. If the IoT devices and IoT edge or gateway are fixed assets that are locked in a fixed location, then you should consider other physical security controls as well to protect direct access to those devices where applicable.

- For securing the IoT cloud or IoT backend:

  i. If your IoT cloud is deployed on the public cloud, then the first thing you should do is to make sure that you have covered the security pillar of the well-architected framework provided by the public cloud provider that you have deployed your IoT backend on. Typically, they provide a massive set of security managed services and security controls that cover all solution security aspects.

ii.   Make sure all the data in transit is encrypted (use TLS/SSL everywhere). This point is always a debatable point between system and solution architects and security architects, since, from a security point of view, having TLS/SSL everywhere is the single security best practice that protects data in transit; however, from a system design perspective, having TLS/SSL everywhere will have some impact on the performance of systems due to computation-intensive processes used for encryption/decryption. This is why some architecture patterns recommend using TLS/SSL for external traffic only and once the traffic enters the internal/private networks, then traffic could be traversed inside the private network as plain traffic (that is, without encryption). Some architects recommend and go with that architecture pattern, since if you have a man-in-the-middle attack inside your private/internal network, then you have a serious security issue. A private network is typically fully secure and protected by a set of firewalls and other security controls. We do recommend using TLS/SSL for external and internal traffic if applicable (that is, provided it will not have a big impact on the performance of the IoT backend components).

iii.   Make sure all the data at rest (that is, in databases, object storage, filesystems, block storage, caching, and so on) is encrypted.

iv.   Use a strong key management and vault solution to store and manage encryption keys and other security credentials used in your IoT solution. If you can (as it is usually expensive), go with a hardware-based key management solution. This will be the best option compared to other software-based key management solutions available on the market.

v.   For IoT solution applications that are accessible over the internet, use ready-made and standard **Web Application Firewall** (**WAF**) solutions that mitigate common web security threats such as SQL injections, **Cross-Site Scripting** (**XSS**), broken authentication and session management, **Cross-Site Request Forgery** (**CSRF**), `https://owasp.org/www-project-top-ten/` for more details on those threats). Some WAF solutions come with advanced features that leverage machine learning in predicting the anomalies and different security threats and offer a mitigation plan as well for such vulnerabilities and anomalies.

vi. Make sure to have a centralized logging and monitoring solution that all your IoT solution components, including IoT devices, IoT Edge, and IoT backend components, integrate with and send their logs and events to that centralized monitoring system. You can use that monitoring solution to check the logs for any security vulnerabilities that have happened or will happen in the future based on the pattern you see in the logs (that is, an anomaly detection feature).

vii. For authorization, use the least privilege principle everywhere. You just give the necessary or required privileges to users or systems to complete their tasks; no wider permissions should be given.

viii. All services and solutions should be accessible only following successful authentication, so use a proper **Identity and Access Management** (**IAM**) system that acts as a single point of authority for your IoT solution. Every component in your solution should be integrated, authenticated, and authorized with that IAM system.

ix. Use modern and widely accepted authentication and authorization standards such as OpenID Connect and OAuth2.0. Use two-factor authentication (or three-factor authentication if applicable) and adopt fine-grain and role-based authorization approaches.

x. Follow the security best practice standards required in the business domain of your IoT solution; for example, if you offer a payment component for your IoT solution (such as paying for a reserved parking lot), then your solution should adhere to the **Payment Card Industry Data Security Standard** (**PCI**). For data privacy, your solution should adhere to the **General Data Protection Regulation** (**GDPR**) and other enforced local and governmental regulations where your IoT solution is deployed and operated.

xi. Have a proper and robust certificate management solution. That solution will manage the life cycle of the X.509 certificate that you typically use as a unique identity for IoT devices.

Monitoring is one of the key enablers for IoT solution security as explained earlier. Therefore, in the next section, let's go through IoT solution monitoring in some detail.

IoT solution monitoring

Monitoring systems are critical components in any large-scale and production-grade IoT solution. You need a centralized monitoring solution (a single pane of glass) that tells you everything you need to know about the different IoT solution operations and performance. Such monitoring systems help in the following:

- IoT solution troubleshooting activities.

- Real-time alerting for issues that occur, which will enable you to quickly respond to those issues that are triggered.

- Proactively identifying the trends and anomalies from the IoT solution logs, traces, and metrics. This is where IoT solution monitoring systems help in securing IoT solutions.

Before we go further with the typical monitoring system architecture that you could use for your IoT solution, we need to first understand the following concepts:

- **Metrics** are data points that can easily be quantified. Metrics are things such as system metrics (that is, CPU, memory, disk utilization), application or business metrics (that is, the number of successful/unsuccessful orders, for example), or custom metrics that you create (for example, counting the number of strings down in the log files and calling that metric the Down metric; then you can apply any threshold rule based on that custom metric value and so on).

  With metrics, you can easily apply alerts and notifications based on the threshold value of those metrics. For example, a monitoring system can send an email to the support team if the average of the CPU metrics is greater than or equal to 80% in a 5-minute window, or create an IT support ticket in the IT support and ticketing system if the number of Down custom metrics is greater than 100 in a 5-minute window.

  With metrics, you use a mathematical operation such as sum or average. Metrics-based monitoring solutions are typically faster and more efficient because they can summarize data using statistics and sampling.

  With metrics, you can also progressively reduce the data resolution and cut the cost of data storage. When data gets older, new data is stored at a 1-second frequency while old data is stored with a 1-hour frequency or less (for example).

  Metrics are great for showing trends over time and can quickly trigger alerts as new data comes in.

- **Logs** are immutable texts that an application generates at a specific timestamp and prints out or appends to a console or file. Broadly speaking, there are three common formats of logs, as follows:

- **Unstructured**: This is the most common, and the oldest, format used by many applications and systems.

- **Structured**: This is the most recent and modern log format. Logs are generated in JSON format.

- **Binary**: This is the most common format that is used mainly in big data and data management solutions. The binary format is usually used when you are looking to increase performance when dealing with log files. They are not human-readable files, but they are machine-readable. Examples of such file formats include AVRO, ORC, and Parquet.

Typically, in an IoT project logging strategy, the project architects or designers define the log format/structure that will be used across all of the IoT solution components. Generally, however (depending on whether a structured or unstructured log format is used), the row or the record of the log in the log file contains a timestamp, the server or the host that printed the log, the source IP, other metadata, and the log message itself, which might contain error code, an error message, and so on.

Log-based monitoring solutions are often slower than metrics-based monitoring solutions because logs provide more granular visibility of what happened in the system and each log record (text) is processed when making queries and calculating trends. In the case of metrics, they are numerical values, so they are much faster at queries and calculating trends.

Log management tools can be built on top of search engines such as Elasticsearch so that they can offer full-text search capabilities.

- **Traces** are a special type of log or a part of log messages that are used to show the trace or the life cycle of an incoming request from when it hit the first server of the application's backend system and kept traversing through other distributed systems until the request was served and the response was returned to the callers or clients of that application. Users often use web clients such as internet browsers to interact with web apps or websites. Requests coming from those browsers or other web clients to a web app backend go through many distributed backend systems that serve that web app. Requests go through systems such as load balancers, web servers, application servers, databases, caching, and message buses.

Traces are typically used to support application performance monitoring. Let's say the response time for a web page of your application is suddenly increased from a few milliseconds to a few seconds. In that case, the support team will investigate why that happened, but the challenge they will have is that they don't know exactly where the problem is or where to start the investigation. Is it in the web servers (that is, server crashes, memory issues, and so on), in application servers (that is, server crashes, memory issues, and so on), in databases (that is, server crashes, memory issues, long-running queries, connection pool issues, and so on), or are issues prevalent in all systems of the application (this is usually a rare occurrence)?

If there is a monitoring solution in place that supports trace monitoring, then you can visualize the journey of the request through all system components, how long it takes in each component or layer, and whether that request gets stuck, causing issues for the component that is serving that request, or whether it is served successfully without any issues.

If there's no tracing, then it will be difficult to figure out quickly where the issue occurred in large and complex distributed systems.

You should consider tracing in your IoT solution logging strategy from the outset. You need to mandate the generation of a unique tracing transaction ID and associate that ID with the request coming from application clients. When the request hits the first backend server of the IoT solution, that first backend server (usually the load balancer, the API gateway, or even a simple basic gateway that sits in front of all your backend systems) generates that unique tracing transaction ID, and then the request should be forwarded to the next system(s) in the chain with the unique tracing transaction ID added to the request object. The receiving system should log the tracing transaction ID as part of its logging and if the receiving system has to call other systems, then the same tracing transaction ID must be forwarded to the next system in the chain, and so on to the end.

Here is an example from real life to better understand the difference between metrics and logs. When you visit the doctor for a routine examination or checkup, the doctor will tell you things such as your blood pressure is X, or your temperature is Y. As you can see, the X and Y are quantities, hence they are metrics. The next question you ask the doctor could be *Doctor, do those values mean I am OK?*. The doctor would reply either with *Yes, those values are normal*, or *No, we need to dig further into why we got value X – for example – so much higher or so much lower than normal.*

The doctor needs to go to the next level of detail, which is checking your *logs* to understand better why the value for blood pressure (let's assume X refers to blood pressure here) is like that. The logs in that case will be things such as you telling the doctor, On day so and so, I ate so and so, and on day so and so, I did my daily exercise for 1 hour, and so on. Now the doctor can understand why the value for blood pressure – your metric – was so high or so low based on your logs.

Now you understand the difference between logs and metrics. The next question is which one should we monitor? Logs only, metrics only, or both? The answer is both (and traces as well). You will need metrics monitoring for quick and real-time monitoring and alerting as querying a metrics repository is much faster than querying a logs repository, and you still need log management as logs give you the detailed information needed for troubleshooting, auditing (to see who did what or who was the source or the trigger of that logging event), and support. You will also need trace monitoring to monitor application performance.

In the next section, we will talk about the typical or generic monitoring system solution reference architecture and explain the most common and well-known monitoring solution on the market, which is the ELK stack (or the modern version, which is called Elastic Stack).

Monitoring system solution reference architecture

The monitoring system solution reference architecture is very simple. Typically, in your IoT solution, you will have many sources of logs and metrics such as IoT devices, IoT edge or IoT gateways, IoT message brokers, IoT rule engines, IoT applications (that is, microservices, APIs, databases, and so on), and infrastructure metrics and logs.

The sources of logs and metrics are on one side; on the other side, or the destination side, is the central monitoring solution where you should store all such logs and metrics data to start the different monitoring activities, troubleshooting, and support for your IoT solution.

Logs and metrics are not sent directly from the source nodes to the central monitoring solution. We have the following components that help in moving logs and metrics from source nodes to the destination nodes:

- **Logs and metrics shippers or beats**: These components are mainly the software agents that are installed on the source node that generate logs or metrics and their job is simply to ship or send the logs and metrics to the destination's central monitoring system.

- **Logs/metrics data processor**: Typically, you will need to process the raw logs and metrics received from the logs and metrics shippers or beats before sending them to the final central monitoring solution system. Logs might come in a different format (that is, if you don't have a logging strategy in place), hence this component's job is to process, filter, enrich, and transform the different log formats into the formats expected by the central monitoring solution system.

Once the logs and metrics are processed, enriched by the logs and metrics processor component, and inserted safely into the central monitoring solution system, the next step will be to access the monitoring solution system and start your monitoring and troubleshooting queries, visualizations, reports, alerting, and log/metric analysis.

There are so many monitoring solutions available on the market – commercial and open source. Some of them focus on metric-based monitoring, others focus on log-based monitoring, and other modern monitoring solutions offer full-fledged monitoring solutions covering all types of monitoring (logs, metrics, and traces).

Dynatrace, AppDynamics, New Relic, Datadog, Splunk, and Elastic Stack are some examples of full-fledged and enterprise-grade monitoring solutions that you could use for your IoT central monitoring solution.

In the next section, we will talk about Elastic Stack as an implementation example of a monitoring system solution reference architecture, which we have just explained. Why Elastic Stack and not other monitoring solutions such as Dynatrace or AppDynamics? We will not answer that question by saying because it is open source, cheap, and so on, which is a valid answer. The main reason for us selecting it for explanation is that it is – based on our experience – the most common and widely used monitoring stack, not only in IoT solutions, but in all other types of IT enterprise solutions.

If you, as an IoT solution architect, have been told to use one or more of those commercial monitoring stacks that we have mentioned (your company might already have licenses of those commercial monitoring solutions), then great – those commercial monitoring solutions are very strong and rich in terms of the monitoring features and capabilities that they provide. Usually, cost is the main concern regarding the use of such commercial monitoring solutions as they are expensive, but if the cost is not your (or your company's) concern, then you should be fine to use them as a backbone for your IoT monitoring solution.

The ELK Stack (and Elastic Stack)

The Elasticsearch, Logstash, and Kibana (ELK) Stack is a group of open-source products from Elastic that form a complete monitoring stack or solution for ingesting logs and metric data from many sources and storing it in the Elasticsearch repository for undertaking searches, analysis, and visualization of that data in real time.

The ELK Stack, as the name suggests, contains the following components:

- **Elasticsearch** is a powerful search and analytics engine that is built on top of the Apache Lucene search engine. Data is stored or indexed in an Elasticsearch cluster in the form of JSON documents, so Elasticsearch is often considered one of the NoSQL data store options.

  Elasticsearch exposes a powerful set of REST APIs that you can use to interact with the Elasticsearch cluster.

  Following are some concepts that you should be familiar with at a very high level to understand how Elasticsearch works:

  - **Document**: This is the basic unit that is stored in the Elasticsearch index (see the Index entity below). It is expressed in a JSON (key/value) pair. This is what the Elasticsearch index does at the end to enable the performance of indexing, search, update, and delete operations.

  - **Index**: This is a collection of documents that have similar characteristics; for example, the customer index contains documents related to different customers (each customer's details are represented in one document).

  There are some other concepts, such as the Cluster, Node, and Shard, in Elasticsearch that it is good to know about and understand, but they are mainly related to the infrastructure or deployment aspects of Elasticsearch services.

- **Logstash** is a free and open-source log management tool or data processing engine that collects logs from different sources in different formats (for example, log4j logs, Apache access logs, Syslog, Windows event logs, and network logs), and then processes, parses, transforms, and enriches such collected logs into a common format that is expected by the destination service. Following that transformation, Logstash streams out the formatted and processed data to its final data store destination (Elasticsearch is one of those destinations in the case of the ELK Stack).

The architecture of Logstash is simple. It is based on the data pipeline processing concept. You configure the Logstash instance on the source node where you need to collect logs by defining pipeline steps in the Logstash configuration file.

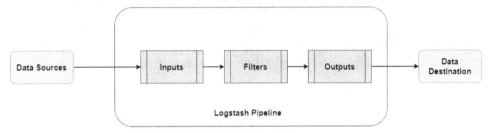

Figure 9.1 – Logstash architecture

The Logstash configuration file has a section for each pipeline's data processing, as per the following:

- **Input plugins**: In this section, you define the source of events/logs that will be read by Logstash. Following are some examples of sources or inputs that Logstash can read from and collect logs from (note that there are more than 50 input plugins for different platforms, databases, and applications):

  - **File**: Logstash reads the logs from a file on the local filesystem where the Logstash instance is installed.

  - **Syslog**: Logstash listens to port 514 for Syslog messages.

  - **Beats**: These are lightweight data shippers that are used to send or forward data from hundreds or thousands of nodes to Elasticsearch directly or to Logstash. There are so many forms of Beats, such as Filebeat, which is used to read and ship logs from files on the source node; Packetbeat, which is used to ship network packets; Metricbeat, which is used to collect and ship metric data from servers and operating systems; Winlogbeat, which is used to ship Windows log events; and Heartbeat, which is used to ship uptime monitoring data.

- **Other input sources**: For example, Redis, Kafka, JDBC, and many other input sources besides.

- **Filter plugins**: These are the pipeline or filter processors that perform data processing on the collected log data (that is, from the input plugins). This is a very important step in the Logstash pipeline as the log data might be unstructured or semi-structured, might contain unnecessary extra information that is not needed to be indexed for searching, or the log data might be missing some other important information that is required during search and troubleshooting; hence, in that step, the filter processor can remove, add, or enrich and format such log data before sending it to the final data store for searching and analytics. This pipeline is optional; that is, you might not need to do any transformation of the collected logs and insert them directly into Elasticsearch. Following are some examples of the filters used in Logstash:

  - `grok`: This type of filter is for parsing arbitrary text and structuring it.

  - `mutate`: For performing mutation tasks such as renaming, removing, and replacing the log data.

  - `Aggregate`

  - `CSV`

  - `JSON`

- **Output plugins**: These plugins or pipelines are responsible for sending the final data (after processing) to its destination. Following are some examples of final destinations that can be used:

  - **Elasticsearch**: This is the destination for the ELK Stack.

  - **Graphite**: This is a popular open source solution for storing and showing metrics in graphs.

  - Others: File, Email, Kafka, HTTP, Nagios, Redis, S3, Stdout, Jira, Hadoop, and so on.

- **Kibana** is an open source, powerful analytical and visualization platform that is designed to work with Elasticsearch. Kibana completes the ELK Stack by offering the visualization part and enabling users to interact with the Elasticsearch store and perform advanced **near-real-time** (**NRT**) searching, build powerful dashboards, perform advanced data analysis, and visualize data in a variety of charts, tables, and maps.

Is the ELK Stack the same as the Elastic Stack or is there a difference between them? The answer is that the **ELK Stack** was the stack that was initially introduced, which has three open source products (that is, Elasticsearch, Logstash, and Kibana). The Elastic Stack also has the same three open source products used in the ELK Stack, plus one additional component added to the ELK Stack, which is **Beats.** Elastic (that is, the company) added Beats to the standard ELK Stack and renamed it the Elastic Stack.

The typical architecture of the Elastic Stack is as follows:

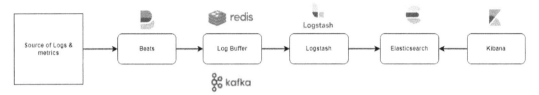

Figure 9.2 – Elastic Stack architecture

So, the main difference between the ELK Stack and the Elastic Stack is the usage of Beats, which raises the following question: Do we really need Beats? In other words, what should we use for log collection and shipment to Elasticsearch: Logstash, Beats, or both?

You need both Beats and Logstash. Beats are so lightweight compared to Logstash instances, hence Beats can be installed on source nodes without impacting the performance of those nodes or taking more compute resources from them. But at the same time, Beats are not designed to execute data transformation or data processing. Logstash, on the other hand, is perfect for doing aggregation and data processing/ transformation. So we recommend using both in your IoT solution monitoring architecture.

Beats can send log data directly into Elasticsearch and bypass Logstash; however, the recommended architecture for large-scale and production-grade IoT solution monitoring systems is to not even send the data directly to Logstash from Beats, but rather send it to a buffer in the middle, such as Redis, Kafka, or RabbitMQ, and then to Logstash, and in the end, Logstash will send it to Elasticsearch (see the architecture in *Figure 9.2*).

The last thing to mention here about the Elastic Stack is that some important features do not come out of the box. For such features, you will need additional extensions to the stack, such as **X-Pack**. X-Pack is an Elastic Stack extension that provides security, alerting, monitoring, reporting, machine learning, and many other capabilities.

Many features in X-Pack are free, such as monitoring, tile maps, Grok Debugger, and Search Profiler, but some other features in X-Pack are paid and require a license (note: this is an enterprise support license (not a software license) as the stack, together with its components, is open source).

Now, let's summarize what you should do as an IoT solution architect to monitor your IoT solution in a few points:

- Make sure that all your solution components, starting from IoT devices up to the IoT cloud/IoT backend, are fully monitored and integrated with a central monitoring solution. Install agents and Beats everywhere to ship logs from the relevant nodes.

- Make sure all component logs and metrics are sent to the centralized monitoring solution.

- Mandate a logging strategy – where applicable – for all components of the solution to cover things such as the following:

  - The format of the logs generated and how tracing will be enforced.

  - The utilization of a proper logging level (TRACE, DEBUG, INFO, NOTICE, WARN, ERROR, FATAL). Choosing a proper logging level will make sure you are not logging too much or too little.

  - No logging for sensitive information, and if there is a need for that, then such information should be anonymized, masked, or encrypted before being stored in a central monitoring log repository.

  - Use logging frameworks that have been tried and tested and that have big community support, such as Log4j, Log4Net, SLF4J, **Apache Commons Logging** (**JCL**), and Log4net, instead of building one from scratch. You should only build a logger wrapper on top of those logging frameworks to avoid vendor lock-in, so you can easily move between those frameworks without breaking the application interfaces.

  - Define the log collection mechanism. Do the logs come as event streams, as explained before, in cloud-native patterns? Or through log collectors/data shippers that are run periodically or on demand on the source application or system? How will you collect the logs of containerized workloads? Will it be through sidecar containers or by collecting the logs from the host server that hosts the containers?

- Define the retention strategy of the log data. For how long will the log data be stored and retained? Especially logs, as logs typically represent such massive amounts of data. You should first check the required retention policy to see what kind of storage will be used. Typically, you would store recent log data (that is, log data that covers the last 30 days, for example) in hot storage, and an expensive solution such as Elasticsearch, and store the old log data in a cold, cheaper solution such as Amazon S3 Glacier or a similar archiving system. The retention policy is usually defined based on regulations (this is most common) or business needs.

- Design and build a highly scalable, distributed, secure, and resilient monitoring solution for your IoT solution that collects logs and metrics from everywhere and enables the operation and support teams to maintain, support, run, and operate such large-scale and production-grade IoT solutions.

Having a centralized logging and monitoring system is a must in running large-scale and production-grade IoT solutions. It will not only help in the supporting and troubleshooting of functional issues, but will also help with the IoT security (as explained earlier), high availability, and resiliency aspects of your IoT solution (that is, by monitoring the infrastructure's and application's key metrics plus real-time notifications and alerting features).

In the next section, let's move on to another of the operational excellence pillars and explain how you can make your IoT solution highly available, resilient, and highly performant.

IoT solution high availability and resiliency

As an IoT solution architect, the first thing you do before starting the design of a highly available and resilient IoT solution is to ask the business stakeholders the following questions:

- What are the expected and accepted **Recovery Point Objectives (RPOs)**?

- What are the expected and accepted **Recovery Time Objectives (RTOs)**?

- What are the expected and accepted **Service-Level Agreements (SLAs)** in terms of system availability and performance?

Let's understand better what RPOs and RTOs are with the help of *Figure 9.3*:

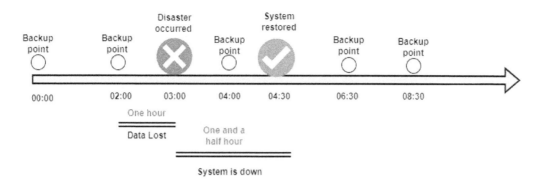

Figure 9.3 – RPO and RTO

Typically, in large-scale and production-grade solutions, there is a backup activity that runs continuously to back up the system data (for example, databases and files) and system software artifacts such as golden virtual machine templates or container images and other artifacts.

Backup is a must-have practice in any large-scale and production-grade IoT solution to restore the system to a working state (that is, a snapshot of a time when the system was up and running) if the system goes down for whatever reason.

Now, with help of *Figure 9.3*, let's assume the business stakeholders told you "*We would expect the IoT solution to be up and running after 2 hours when a disaster happens and hits the IoT solution's systems, and we would expect or agree to lose 2 hours of system data*".

With that example in mind, the RPO is equal to 2 hours, and the RTO is equal to 2 hours.

In *Figure 9.3*, there are backup points (or backup snapshots) that are taken every 2 hours. The last backup taken was at 02:00 a.m. That backup backed up the data entered into the system between 00:00 and 02:00 a.m., and then at 03:00 a.m., a disaster hit the system, so any data entered between 02:00 a.m. and 03:00 a.m. will be lost (assuming a worst-case disaster scenario) as the next backup was supposed to run at 04:00 a.m. According to the RPO agreed with business stakeholders, losing 2 hours of data is OK, so we are fine as we are losing less than 2 hours of data, in this case, 1 hour's worth of data.

The story continues. The support and operations teams work very hard after the d isaster to restore the system so it's up and running again. They manage to restore it – with the help of backup snapshots and so on – by 04:30 a.m. (one and a half hours after the disaster occurred at 03:00 a.m.). According to the agreed RTO with the business stakeholders – that the IoT solution's systems should be restored within 2 hours – we are fine as the systems have been restored in under 2 hours; that is, they were restored in one and a half hours.

Note, we have mentioned that we are assuming our IoT solution's systems were hit by a **worst-case scenario** disaster such as a flood or earthquake. In other words, we are assuming here the types of disaster that would have a severe impact on the data center's facilities and might corrupt or ruin the storage media used in storing system data; otherwise, the systems might be down from only the compute point of view. However, the data that was entered before the disaster occurred is stored durably and safely in the database or any other storage medium that is used as part of the solution. Hence, in that case, there's no loss of data.

Knowing the RTO and RPO values is critical input that you must collect from the business stakeholders to define your IoT solution's recovery strategy; for example, if the RPO and/ or RTO values are aggressive, such as the values for RTO and RPO being in minutes, then the time of the backup snapshots must be equal to or less than the required RPO, and the process of restoring the systems must be fully automated to restore the systems in less than or equal to the required aggressive RTO. We will explain this in greater detail later.

Now, regarding the **SLA** of the IoT solution's system availability, you need first to understand what system availability means. It means that any time the system's clients (any client) send a request to the system, the client must get a response to that request immediately. There is a mathematical way or science in calculating the system's availability time, but to keep it short, the following matrix (note: the numbers in the table are approximate) shows you the different levels of availability and what they mean:

System downtime per day	System downtime per year	Uptime percentage	Comment
2.4 hours	36.5 days	90%	If the system has such metrics, then the system availability SLA is 90% – in other words, one nine.
14 min	3.65 days	99%	Two nines
66 sec	8.76 hours	99.9%	Three nines
8.6 sec	52.6 min	99.99%	Four nines
0.86 sec	5.25 min	99.999%	Five nines
8.6 msec	31.5 sec	99.9999%	6 nines

From our practical experience, an availability SLA of up to 4 nines (99.99%) is an achievable SLA; however, it is challenging and not easy to achieve. To achieve it, you must design your IoT solution's systems very well, as we will explain later.

Beyond the 4 nines SLA, there are really such aggressive SLAs that not all companies will be able to offer them, that is, 5, 6, 7 nines, or more. We have seen companies say that their solutions or systems have been designed for 5 (or 6 or more) nines SLA availability. The tricky words here are **designed for** as they are not contractually binding for the SLA. So, you might have an actual system availability SLA of 99.9% despite the company saying **designed for 5 nines**. In that case, the company is not legally required to pay you any credit or compensation for that simply because they are not really breaking the SLA agreed with you (that is, in the contract, they didn't agree to an SLA of 5 nines; they just wrote designed for a 5 nines SLA.

Another concept to understand is **resiliency**. Resiliency means your solution or systems can continue serving the requests they receive even if they are encountering some faults or issues.

We all know faults or errors will happen anyway, but the most important thing you should consider in the design of an IoT solution is how to react to such faults and ensure the system continues working without being impacted or going down.

Performance is another aspect that you should consider for your IoT solutions, and it means systems can react quickly to the requests they receive without any further delay than the expected latency of the system. In highly performant solutions, the response typically comes back in under a second, or it might take more than a second in some cases (that is, the response comes back within 1 or more seconds) if there's lots of processing required in the backend.

Redundancy of computing resources is one of the key enablers in achieving non-functional aspects such as high availability and resiliency (or some of them), but it usually comes with additional costs, such as these:

- To have your solution Geo-R (Geographically Redundant), then you must have your solution deployed in at least two different data centers in two different geographical regions.

  No doubt this is the best solution for business continuity and disaster recovery in case a disaster occurs in a complete region (rare, but it could happen) or the business has some regulation, such as a data locality regulation that forces the business to have customer data stored in a specific region. Having said that, this option is also the most expensive compared to other redundancy options we will mention next.

- Another level of redundancy that might be accepted is local (Local-R) or single-region redundancy, which means that in a single region, you might deploy your IoT workload in two different data centers within that region.

Cost-wise, this option might be the same as the Geo-R option if you use the same number of data centers as in the Geo-R option. The only real difference between those two redundancy models will be the data centers' locations (in the case of Geo-R, the two data centers are in two different regions, while in Local-R, the data centers are within the same region). From a risk point of view, the Geo-R option is much better than the Local-R option for business continuity plans.

- Another level of redundancy could be on a single region and a single data center (this is a very basic setup and usually not for large-scale and production-grade IoT workloads). In that option, you might use different racks, power supplies, and so on, in the same data center. It could be that you have one rack on the first floor of the data center facility and the other rack might be in the same building (on another floor) or even in another building that is nearby.

The devil is in the details. We could continue listing other different redundancy options that we can use for our IoT solutions, but in general, the options we have explained are the most common and most frequently used redundancy options.

If your IoT workload is in the public cloud, then there is more flexibility and many redundancy options. For example, in a single region of a public cloud provider, you have multiple availability zones, and each availability zone has multiple data centers. Each data center has many fault and update domains, hence a single-region system in the public cloud might be deployed and run – behind the scenes – in more than four data centers in different locations within that region.

Now it is time to see what you could do in the IoT solution design to make it highly available, resilient, and performant:

- **On the IoT device level**, you don't have much to do, but here are some points that you should consider for making the IoT device more resilient, reliable, and highly available:

  - Make sure IoT devices can work and support offline mode as the connectivity will drop at some point or become unreliable, so IoT devices should continue working even if they are disconnected from the internet. To support the offline mode feature, the IoT devices should be able to do some processing and store some data locally while they are disconnected from the internet. If the IoT devices are low-power and resource-constrained, then you might handle any local computational and storage requirements in the next powerful, capable level, which is typically the IoT gateway or the IoT Edge.

- Make sure IoT devices support features such as digital twins/device shadowing and that there is a sync process between the device and its twin in the cloud when the connectivity is back.

- Protect IoT devices – as explained earlier – from a security point of view, to avoid DDoS attacks on those IoT devices. A DDoS attack could kill the batteries of those devices by just keeping the IoT devices (which are usually low-power) ON all the time, hence the batteries will be fully drained and the devices will become useless and will not work.

- Make sure to support dual-connectivity options either in IoT devices or in the IoT Gateway/Edge. If one of the device connectivity options drops, the IoT device will fail over and use the other available connectivity option, something like your mobile handset connectivity options, which typically include Wi-Fi and mobile broadband connectivity. If the Wi-Fi drops, or vice versa, the other connectivity option will pick up and continue the connectivity.

- Make sure to monitor device metrics such as battery percentage, battery voltage, network packets in and out, TCP connections, connection rate to the IoT cloud, and other metrics that could help in predicting any maintenance required before the device completely shuts down and is out of service.

- Make sure to monitor the software running inside IoT devices to check software bugs, crashes, security vulnerabilities, and other custom metrics that you build as part of your IoT device firmware or software. If there's any patch required, then you can do that immediately using the over-the-air update feature of the IoT device management solution, as explained earlier in the book.

- **On the IoT Edge level**, if it is an industrial IoT solution, you are most probably leveraging some sort of IoT Edge capabilities, at least using a powerful IoT Gateway device that connects to the internet from one side and to the IoT devices' endpoints from the other side.

IoT Gateway or IoT Edge is your last chance to have reliable, highly available, and resilient IoT devices on the ground (that is, away from the IoT cloud), hence all of what we explained before about IoT devices' resiliency and availability should be applied to the IoT Gateway device too, with some additional checks related to IoT Gateway devices as IoT Gateway devices are usually more powerful devices compared to endpoint IoT devices. In other words, they (IoT Gateway/Edge devices) might be treated – from a security point of view – as a typical backend server.

- **On the IoT cloud/backend level** is where you apply the same design principles and practices you typically apply for standard IT workloads running on the cloud. We will mention some of the design principles that you should use and apply to IoT cloud or IoT backend applications and solutions:

  - Try as much as possible to use managed services in your IoT backend components where applicable. We know that option might seem expensive compared to the DIY (do it yourself) option, but from our experience, we find that using managed services might cost the same as, or be cheaper than, building, supporting, and maintaining applications that are built in-house. To clarify further, in your IoT data analytics solution, for example, you could say, we will use open source Hadoop and its massive ecosystems, and then you would look around in your organization for highly qualified resources who have the required skills to run and operate a Hadoop cluster(s) and other related technologies. You will be lucky if you find resources with such advanced skills easily. You might have to hire those kinds of highly skilled resources to do that job. On the other hand, if you go with a Hadoop managed service from a service provider such as Amazon, Azure, or Google, then you will have a Hadoop cluster up and running in a few minutes. You will not need to worry about the operational aspects (installing, patching, and so on) of running and supporting that Hadoop cluster; it is the service provider's responsibility, and you will realize the cost per year of such a managed service is equal to the annual salaries that you would pay for highly qualified resources that you would otherwise have to hire. (The same concept (managed versus DIY) can be applied to Kubernetes, databases, Elasticsearch, and so on.) In short, managed services are (and must be) always highly available, scalable, and resilient, so your IoT solution will inherit such benefits (high availability, scalability, and resiliency) from those managed services.

  - In any layer of your IoT cloud, use redundancy of resources (compute resources) everywhere. Always go with at least two nodes and a load balancer in front of them. Use an autoscaling mechanism to automatically scale the workload based on standard metrics such as CPU or memory usage, or custom metrics such as the number of messages in a queue.

  - For resiliency and scalability, make sure to build your IoT backend components as cloud-native and 100% stateless to enable the horizontal scalability of your IoT cloud/backend.

  - If your IoT backend workload is containerized, then use a container orchestration engine such as Kubernetes to gain out-of-the-box resiliency (for example, self-healing) and scalability (for example, pod autoscaling) options that come built -in with Kubernetes.

- Always have your IoT backend deployed in at least two data centers (the same region or preferably different regions for Geo-R). Make sure that your IoT backend is locally fully redundant in a single region.

- For an IoT solution disaster recovery and business continuity strategy, you should go with a Geo-R deployment model (see the Geo-R deployment models section for further information).

Geo-R deployment models

Based on the RPO, RTO, and SLA values you collected from business stakeholders and the cost or budget allocated for the IoT solution, you can choose any of the following Geo-R deployment models for an IoT solution.

Backup and restore

If the RTO and RPO values are in hours or the allocated cost/budget is small, then the backup and restore Geo-R model should be your chosen model.

In that case, you have limited allocated costs/budget from the business for disaster recovery and the business continuity plans, but the good news is that you have been given higher RTO and RPO values, which gives you some more flexibility and freedom in choosing longer backup snapshots, that is, every 2 hours or even more, and flexibility in the system recovery or system restoration process too, that is, performing some restoration tasks manually if needed as you have enough time (a high RTO) to restore the service.

In that model, the story starts when you have – at least – two data centers that are used in hosting your IoT workload. One of those data centers acts as the primary data center or site, with all traffic going to that data center, as shown in *Figure 9.4*. The secondary site might have the IoT workload in a shutdown state (that is, ready to start when needed) or might not have any IoT workload yet (during the restoration process, the IoT workload will be deployed first and then the restoring process of the system will be triggered next).

There is no production traffic (yet – before a disaster) to the secondary site as long as the primary site is healthy/active and serving production traffic. There is a backup process run in the primary site to back up data from different data stores used in the IoT solution, such as databases (SQL, NoSQL), object storage, filesystems, and caches.

For the compute part, which typically includes resources such as a **load balancer (LB)**, VMs, containers, and functions (FaaS), there are so many options that could be used to back up and restore the compute services. Here are some examples:

- If your IoT backend workloads have been designed to be cloud-native and 100% stateless, then you can simply have the Infrastructure as a Code and software configuration scripts (that is, Terraform, Ansible, Puppet, and so on) stored in a source code repository such as Git, and then you can rerun such scripts anywhere and launch the compute services of your IoT workload.

- You might use what is known as VM Golden Images. Such VM images have the VM operating system plus additional software and configurations you install and configure in those VMs. This option might be used for quick restoration of the compute services. It also can be used with cloud-native software (that is, 100% stateless) for the same good reason (quick restoration).

IoT workload software artifacts (for example, VM templates or images, container images, software binaries, and executables) are typically stored in an artifact repository.

That artifact repository is also used during the restoration process alongside the backup repository:

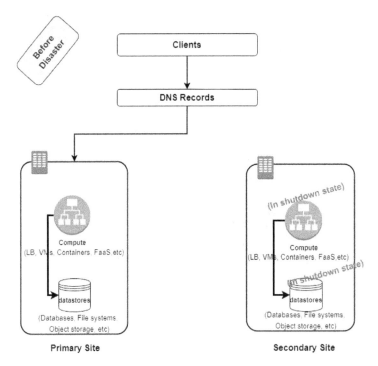

Figure 9.4 – Backup and restore Geo-R deployment option – before a disaster

Now, if a disaster hits the primary site and it goes down completely (or part of it – depending on your architecture decision, it might be that even if a part of it goes down, such as compute or data stores, then it fails over to the secondary site), then the restoration process will be initiated, as shown in *Figure 9.5*:

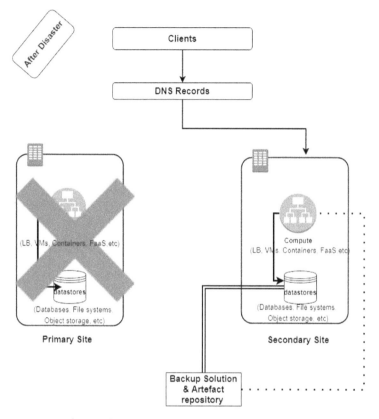

Figure 9.5 – Backup and restore Geo-R deployment option – following a disaster

The restoration process is quite easy. You need to launch or create from scratch in the secondary site the compute services of the IoT solution and launch or create the data stores of the IoT solution as well. Then, restore the data from the recent backup snapshot taken from the primary site before the failure, and finally, change the DNS record to refer to the secondary site entry point. With that, all traffic will be shifted from the primary site (that is, the failed site) to the secondary site.

Pilot light

If RTO and RPO values are in a high number of minutes (for example, 50, 40, or 30 min) or the allocated cost/budget is medium, then the **Pilot Light** Geo-R model should be your chosen model.

In that case, with such given inputs, you have quite aggressive RTO and RPO values, but the good news is that the cost allocated is not small, hence the best model you should use in that case is Pilot Light, which means you have your IoT workloads deployed in two data centers. One data center is fully active while, in the other data center, no deployed resources are running, except for services or resources that depend on the data, such as databases, object storage, and data volumes, as shown in *Figure 9.6*:

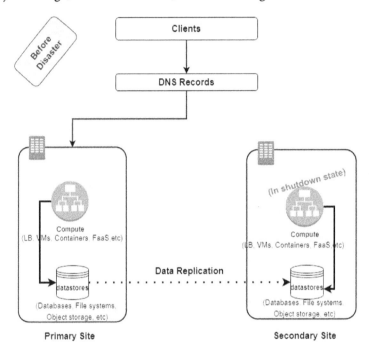

Figure 9.6 – Pilot Light Geo-R deployment option – before a disaster

Since the RTO and RPO are a bit aggressive, you can't count on the backup process with 2-hour or 1-hour snapshots, especially if you have a large data volume. In such cases, data replication and data mirroring techniques are the best options for backing up the data from a primary site to a secondary site.

There are many tools and solutions for data replication and data mirroring between data centers or sites and different synchronization options, such as synchronization data replication (often used if the sites are within the same region) and the asynchronization option. This is the most commonly used and preferable option that is used in data replication, regardless of whether the sites are within the same region or in different regions. The only concern with that option – compared to the synchronization option – is latency, as the data is replicated instantly between the sites in the synchronization data option, whereas in the asynchronization option, there is usually a time gap between the data being replicated between the primary and secondary sites.)

Now, if a disaster hits the primary site and it goes down completely (or part of it – depending on your architecture decision, it might be that even if a part of it goes down, such as compute or data stores, then it will fail over to the secondary site), then the restoration process will be initiated, as demonstrated in *Figure 9.7*:

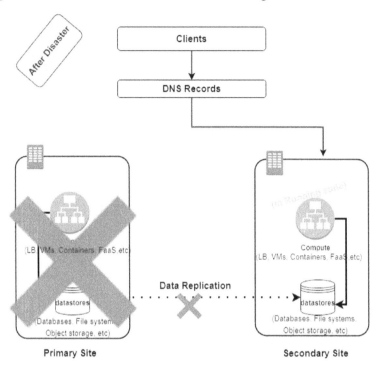

Figure 9.7 – Pilot Light Geo-R deployment option – following a disaster

The restoration process in Pilot Light is also quite easy. The compute resources will start to be moved from a shutdown state to a running state. The data stores already have a copy of the data (thanks to the data replication process that was running between the primary site and the secondary site). Finally, you just need to change the DNS records to shift the traffic from the primary site (that is, the failed site) to the secondary site.

This model is called *Pilot Light* as some services (mainly data services) are kept running in the secondary site to be used as an ignition for launching the IoT workload quickly in the secondary site when disaster hits the primary site.

Warm standby

If the RTO and RPO values are in a low number of minutes (for example, 20, 10, or 5 min), or the allocated cost is high, then the **Warm Standby** Geo-R model should be your chosen model.

In that case, you have an aggressive RTO and RPO, but you have the budget required to achieve those RTO and RPO values, hence the model you should use in that case is **Warm Standby**, which means you have the secondary site workload up and running (on a small scale) but not serving real or production traffic, the primary site is up and running and serving real or production traffic, and data synchronization between the primary and secondary sites is running as shown in *Figure 9.8*:

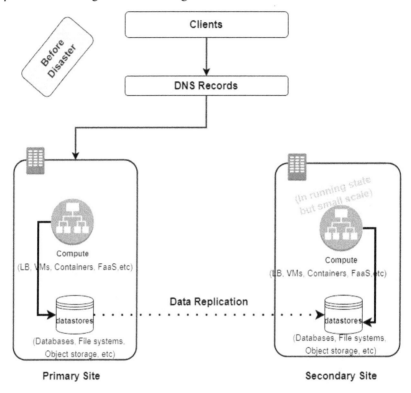

Figure 9.8 – Warm Standby Geo-R deployment option – before a disaster

The main difference between the Pilot Light option and Warm Standby is the fact that the compute resources in the case of the Warm Standby model are running and warm (ready to serve the traffic) but on a small scale (that is, to save some money as there's no real traffic yet), while in the case of the Pilot Light option, only data services are running and in the restoration process, you start or create the compute services.

Now, if a disaster hits the primary site and it goes down completely (or part of it – depending on your architecture decision, it might be that even if a part of it goes down, such as compute or data stores, then it fails over to the secondary site), then the restoration process will be initiated, as shown in *Figure 9.9*:

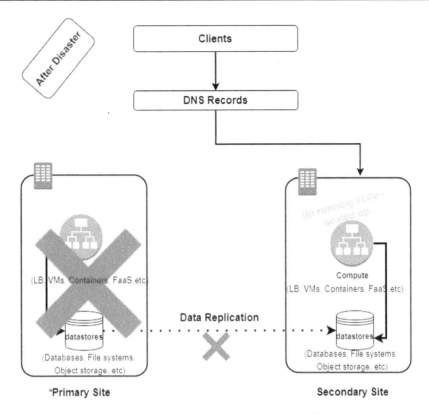

Figure 9.9 – Warm Standby Geo-R deployment option – following a disaster

The restoration process in Warm Standby is done through scaling up the compute resources that are in a warm state, either manually or automatically (the recommended and preferable option). Ideally, you should leverage features such as an autoscaling group with an LB in front of it to scale up the compute resources automatically and quickly.

Active-active

If the RTO and RPO values are in seconds or less, or the allocated cost/budget is high, then the **Active-Active** Geo-R model should be your chosen model.

In that case, you have a very aggressive RTO and RPO, or you might not have any RTO or RPO at all. In other words, the business team expects the IoT solution to not go down at all and no data should be lost. These are very tough requirements, but the business team allocates a very high budget to achieve them.

The **Active-Active** Geo-R deployment model means that you have your IoT solution fully deployed in two (or more) different data centers in two (or more) different regions. The two sites are production-grade and ready to serve production traffic, as shown in *Figure 9.10*:

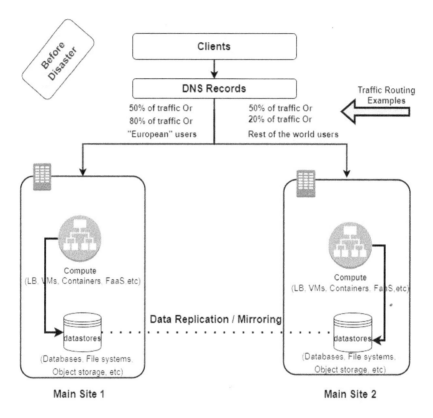

Figure 9.10 – Active-Active Geo-R deployment option – before a disaster

The traffic routing strategy in that model is very important. You can define the routing strategy as 50% of production traffic going to Main Site 1 (note, in Active-Active, both sites are considered the main site), with the other 50% going to Main Site 2, or 80% versus 20%, or whatever values fit your business model. You might also split the traffic based on the location or geographical area of end users of the IoT solution; for example, traffic coming from European users goes to Main Site 1, while traffic coming from the rest of the world's users goes to Main Site 2 or vice versa (those are just examples).

If a disaster hits either of the sites and the impacted site goes down completely (or part of it – depending on your architecture decision, it might be that even if a part of it goes down, such as compute or data stores, then it will fail over to the secondary site), then the restoration process will be initiated, as indicated in *Figure 9.11*:

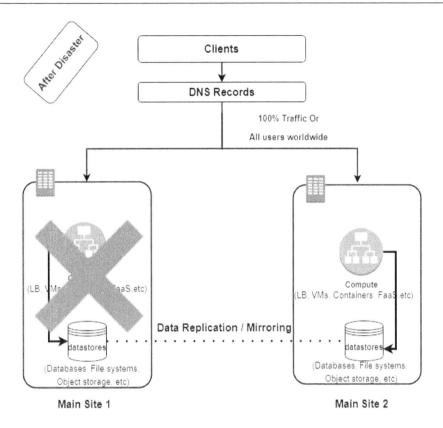

Figure 9.11 – Active-Active Geo-R deployment option – following a disaster

The restoration process in the **Active-Active** model is done automatically on the DNS level. In other words, only the DNS records need to be changed to refer to the new main site. That change to the DNS records must be done automatically. Typically, a modern DNS service has a health check endpoint configured for each of the two sites and if the health check endpoint of a site goes down or becomes unhealthy, then the DNS record will automatically fail over to the other healthy site. Note that the changing of the DNS record (from the old IP (unhealthy site) to a new IP (healthy site)) is done instantly, but you still need to make sure the DNS caching **time to live (TTL)** value of that record is set to 0 to disable the caching of that DNS record on the client side. You don't need a situation where you perfectly fail over on the server side but the clients keep calling the old DNS record value (the old IP of the failed/unhealthy site). By disabling DNS caching, the clients will get a fresh response from DNS services every time they try to access the IoT solution.

When it comes to disaster recovery options, there are lots of details and design considerations. We have tried as much as we can to make it generic. If your IoT workload is hosted and run by a public cloud provider such as Amazon, Azure, and Google, then you should follow the detailed architecture, design, and best practices provided by those public cloud providers for implementing disaster recovery and business continuity plans. Let's take **AWS** as an example:

- With a click of a button, script, or API, you can create a local-redundant solution by deploying the workload into multiple availability zones in a single region, or you can create a geo-redundant solution by deploying into multiple geographical regions that also have local redundancy (availability zone redundancy).

- For the database (relational database), the data synchronization process between the database master and slaves is typically a complex process, but with a managed service such as AWS RDS, you don't need to worry about that complex process. As for the **local-redundancy** model, RDS supports what is called the Multi-AZ model (master and slave databases are in the same region and the data synchronization is done instantly). For the **Geo-R** model, you can have an RDS **Read Replica** (a read-only database) in the secondary site (in a different region from the main site) and it will be synced from the master database in the main site asynchronously. If a disaster befalls an entire region, then the read replica can be promoted to the master database (receiving read and write traffic) or you can go with fully managed database options such as AWS RDS global database or Amazon Aurora.

- For NoSQL databases, AWS DynamoDB provides what is known as global tables that help a lot in Geo-R cases as the synchronization (in different regions) is done under the hood by AWS.

- For the object storage service, Amazon S3 has a feature called cross-region replication that helps in Geo-R cases and other cases as well, such as data compliance/data locality cases.

- For VM-based workloads, AWS provides services such as CloudFormation, which is basically an implementation of the Infrastructure as a Code concept. In the CloudFormation template, you define the VM images or Amazon Machine Images (AMIs) per region and you can have a mapping section in that template to choose the right AMI in a specific region. With a click of a button in the AWS console or through the command line or API, you run the template and you will have the compute infrastructure (EC2 machines, load balancer, autoscaling, and so on) provisioned, up and running (that is, assuming golden images or software configuration management scripts such as Ansible and Puppet are used with CloudFormation), and ready to serve traffic.

- For container-based workloads, AWS provides services such as **Elastic Container Registry** (**ECR**), which has a cross-region replication feature that helps in Geo-R cases (the container images will be stored in different regions). For Kubernetes workloads that run those container images, you can use CloudFormation (or Terraform) to launch the cluster in any region and connect that cluster to the ECR in that region. To deploy containers in Kubernetes, you can go with native Kubernetes tools or use things such as Helm Charts.

- For DNS, AWS provides **AWS Route 53**, which is a great DNS service (the AWS availability SLA for that service is 100%, which means it never fails to respond to any DNS queries). Using a service such as AWS Route 53 will enable you to define the routing policy between your regions easily and smoothly in AWS (that is, of a perfect fit and well-integrated) or even outside AWS (private clouds or data centers).

- Other **AWS managed services** that you might use in your IoT workload include **AWS IoT Core** and **IoT Analytics**. Those services are natively resilient, redundant, scalable, and highly available.

We could continue talking about the different AWS services and how they perfectly help in building highly available, resilient, and performant IoT solutions, but we know you might already use or be thinking about using different public cloud providers for your IoT solutions. So, as explained earlier, we've tried to keep the disaster recovery options and strategy more generic, but at the same time, we've tried to give you a hint of how easy and smooth it is to implement them in public clouds.

We've come to the end of the third (high availability) and fourth (resiliency) operational excellence pillars. In the next section, we will cover another important pillar or aspect of IoT solution operational excellence, which is **automation**.

IoT solution automation and DevOps

To understand why **automation** is such an important aspect of best-in-class IoT solutions, think about the following: you have designed and built a highly available, scalable, resilient, and fully monitored IoT solution. One day, you get a notification or an alert from the IoT monitoring solution that there's a software bug or a security vulnerability discovered on the IoT device, IoT Edge, or in any IoT backend component. Now, if you don't have an automated process and tools in place to enable you to quickly release a software patch, then you lose all the benefits you get from a system like an IoT monitoring system. What is the point of being notified about issues if it takes such a long time to fix such issues?! Automation is what closes the circle of having a good IoT solution with high operational excellence.

Delays in releasing software patches to your IoT devices or the IoT cloud could have severe impacts on the company's brand, revenue, and customers. The question is, What could delay a software release?

In the traditional software release process, an idea (that is, a business requirement) or software bugs get assigned to the development team manually or through an automated channel such as Slack or a software management tool such as Jira. The development team will look into the new features or bugs and start the software coding, do unit testing, and integrate them with other components in the staging or quality assurance environment. Then, the development job is partially completed, and the testing team will start their next job.

The testing team will run their testing scripts manually or automatically. If there are issues (bugs) discovered, they will assign them back to the development team to fix, and so it goes until the testing process is completed. Then the code is ready to be deployed in the production environment. At that time, the development team's job is complete and it is the operations team's responsibility to deploy and maintain the code in the production environment.

Once the code is deployed in production, the operations team will keep monitoring the solution, and if there are any issues (bugs) discovered, they will assign them to the development team, or if the business team wants to release new features, they will assign that to the development team, hence the software release process will start again to cover either bug fixes or new business features.

In the traditional software release process, delays happen because of the following:

- **Miscommunication and lack of collaboration**: As explained, there are so many teams involved in the software release process with different backgrounds. The development team focuses a lot on software development and they might not be interested in knowing about the infrastructure part and how their software gets deployed and run in production environments. The operations team, on the other hand, doesn't focus too much on software development and the differences between different software frameworks and stacks that are used in software development. This brings lots of miscommunication and misunderstanding between the teams, which, in the end, means more time spent on releasing the software.

- **Conflicts of interest**: In the traditional software release process, there are many conflicts of interest between different teams. The most famous conflict of interest is the one between the development team and the operations team. If you are on the development side, then your goal as a developer is to release the new software feature or bug fix quickly, while if you are on the operations side, then your goal is to maintain the system stability and reliability as the production environment is not like the development environment. Operations teams – naturally – are usually reluctant to make any changes to systems in production environments. Such conflicts cause delays in the software release process.

- **Lots of manual work**: Manual work, by default, slows down the software release process and is error-prone. So, the more the different teams do their work manually, the more the whole software release process will be impacted and delayed. For example, on the testing side, if the testing team runs test scripts manually, then they will take a long time to execute. On the operation side, if the operations team deploys manually (that is, prepares environments manually, configures Jenkins build jobs manually, runs deployment scripts manually, and so on), then it will take a long time to deploy the software in a production environment. The same applies to the security side (that is, if the security team does security checks manually, and so on), not to mention that the rollback and recovery scenarios with such manual work are difficult and complex, and typically will take a long time to bring the systems back to their previous working state.

    Delivering fast is the key here. This is why an IoT solution blueprint should include a recommended software delivery process and automation tools that will help in releasing the software of the different IoT solution components fast and with high quality.

We finally come to **DevOps**. What is DevOps and why should we use it in our software release process? As for Why?, we already answered that question: simply, DevOps involves streamlining the software release process. In other words, DevOps is all about releasing high-quality software products quickly and efficiently.

For *What*?, you need first to understand from where the name came. It is obvious that the first part of the name came from Development (Dev) and the second part came from Operations (Ops). With DevOps, there is no separation or isolation between the development team and the operations team. In DevOps, the main principle is that you build it and then you run it.

With DevOps principles, new roles or jobs started to appear on the IT market, such as DevOps engineer and DevOps architect. On that point, we started to see some developers move toward the DevOps engineer role by counting on their development background and learning and adding to their development skills other operations and system administration skills, such as cloud computing, networking, security, continuous monitoring, container technologies, package management, Infrastructure as a Code, CI/CD pipelines, Kubernetes, scripting, and operating system skills. We also saw operations teams move a bit toward the software development world to be able to fill DevOps roles by adding some software development skills to their operations and system administration skills (that is, understanding at a high level different software architecture paradigms such as three-tier architecture and microservices; different ways to build software using different tools such as Maven, Gradle, and npm; and different software artifacts that are deployed in the production environment, such as JARs, Docker images, EARs, and WARs).

To some people, though, DevOps is just a set of tools that help in building what is known as **CI/CD (Continuous Integration/Continuous Deployment or Delivery)**. This is partially right, but DevOps is a combination of a set of principles or philosophies and a set of tools to streamline and automate the software release process.

Configuring CI/CD pipelines to run continuously is the main role of the DevOps engineer. This is why the famous DevOps logo shows an infinite cycle of software release activities or stages, as shown in *Figure 9.12*:

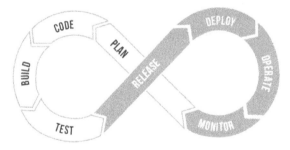

Figure 9.12 – DevOps (image source: https://jfrog.com/devops-tools/what-is-devops/)

For each stage or activity, there are many tools that you could use. You can build CI/CD pipelines from scratch using open source or commercial tools, or you can go with a cloud provider DevOps suite such as the Microsoft Azure DevOps service , AWS CI/CD services that contain different AWS CI/CD Bui such as **AWS CodeCommit** (a fully managed source control source control service that hosts secure Git-based repositories, CodeCommit makes it easy for teams to collaborate on code in a secure and highly scalable ecosystem), **AWS CodeBuild** (a fully managed continuous integration service that compiles source code, runs tests, and produces software packages that are ready to deploy on a dynamically created build server), **AWS CodeDeploy** (a fully managed deployment service that automates software deployment services, such as Amazon EC2, AWS FargateAWS Lambda AWS Fargate, AWS Lambda, and on-p and **AWS CodePipeline** (a fully managed continuous delivery continuous delivery service that helps you automate your release pipelines for fast and reliable application and infrastructure updates such as Jenkins).

From our practical experience, we have found that using cloud providers' managed CI/CD pipeline services is more efficient, fast, and reliable in comparison with hosting and running complete CI/CD pipeline tools and infrastructure ourselves. IoT solution operations teams should not be overloaded with the additional overhead of maintaining and supporting DevOps pipeline tools as they are typically a massive set of tools, including the following, to mention a few:

- For **requirement management** and **issue and project tracking**, you might use tools such as Jira, Assembla, and Axosoft.

- For **code repositories and commits**, you might use tools such as Git, GitHub, GitLab, and Bitbucket.

- For **code building and packaging**, you might use Maven, Gradle, npm, and Docker.

- For **code testing and scanning**, you might use tools such as SonarQube, SOASTA, Snyk (container vulnerability scanning), Anchore (scanning container images for many different scanning aspects, including security vulnerabilities, open source licenses, and so on), OWASP Dependency-Check, PHPStan, OWASP Zap (DAST), Coverity (static code analysis), Junit, NUnit, Jmeter, Selenium, black duck, cucumber, Gremlin, and BlazeMeter.

- For **deployment**, you might use tools such as Jenkins, GitLab, Bamboo, Spinnaker, and Octopus Deploy.

- For **configuration management**, you might use tools such as Chef, Puppet, and Ansible.

- For a **CI/CD pipeline orchestrator**, you might use tools such as Jenkins, GitLab, CircleCI, Bamboo, and JFrog.

- For **software artifacts**, you might use tools such as Nexus and Artifactory.

Another concept that is used in the DevOps CI/CD pipeline that you should be familiar with is the different deployment strategies that you can use during the deployment activity. Typically, you have the old code or software of the IoT solution that is running in the data centers or in the cloud, and you have new code or software that should replace the old code. What deployment method or strategy should we use to do that? There are four common deployment methods:

- **Big-bang deployment**: In this model, you deploy the new code to all the servers that host the old code at the same time. This is a risky and very old approach that is not used in modern DevOps practices.

- **Rolling deployment**: In this model, you undertake deployment step by step by rolling or updating some of the servers with new code and keeping the other servers with the old code. Then, you continue updating other servers that have the old code till all the servers have been updated with the new code. This model is also a bit risky. The rollback is a bit tricky in terms of keeping track of which server has the old code version and which server has the new code version, and it might take some time to be done correctly.

- **Blue-green deployment**: This is the most used deployment method nowadays. You simply have one environment that hosts the old code version and you have another environment that hosts the new code version. After testing and all the verifications required for the new code version in the new environment are done, you simply shift the user's traffic to the new environment with the new software version.

- **Canary deployment**: The idea of canary deployment is to give a few users access to the new software version. After you collect feedback from those few users and apply any necessary changes, you can then move all users to the new version of the software.

Unfortunately, we cannot cover such exciting topics as DevOps in detail as they are out of scope for this book, but we have scratched the surface to show IoT solution architects and designers of large-scale and production-grade IoT solutions how important those topics are and how they help in completing the perfect picture of a production-grade E2E IoT solution.

Some other hot topics emerged from DevOps, such as DevSecOps (that is, adding security automation to the process), DataOps (that is, an automation process for the data analytics domain), AIOps (that is, using AI techniques, machine learning, and **natural language processing** (**NLP**) to automate IT operation processes), and even NoOps (that is, everything is automated and there's no need for human intervention in the process). It is worth reading further about such topics if you are interested.

Summary

In this chapter, you have learned about different non-functional aspects of IoT solutions such as security, monitoring, high availability, reliability, resiliency, and automation.

We have covered each aspect of such non-functional or operational excellence aspects in different IoT solution layers (IoT devices, IoT edge, communication, and the IoT cloud).

In the next and final chapter of this book, we will recap and summarize what you have learned by going through some of the E2E IoT solution reference architectures provided by different hyperscale cloud providers.

10
Wrapping Up and Final Thoughts

Here, we've come to the last destination of our IoT journey in this book. In the previous chapters, we have covered lots of business and technical details about designing, building, and operating large-scale and production-grade IoT solutions. We went on an exciting journey with many IoT stations and now we've reached the last IoT station on our journey.

The first IoT station on our journey was *Chapter 1, Introduction to the IoT – The Big Picture*, which gave you a holistic view of IoT in general and covered different business and technical aspects of large-scale IoT solutions. We covered the answers to three common questions that are typically asked before proceeding with any new business idea or technology. We also covered the *What?*, *Why?*, and *How?* questions of IoT.

Chapter 1 Quick Message

We covered so many things in *Chapter 1, Introduction to the IoT – The Big Picture*, but if there is one important and quick message we would like to share with you from that chapter of the many messages we highlighted, it would be that IoT technologies – and other technologies – always come along to solve business problems. So before you start the design and solution architecture of the IoT solution, you must understand the business requirements or the business problems and understand how IoT technologies will solve those business problems.

Then, the IoT train moved on to the next IoT station, which was *Chapter 2, The "I" in IoT – IoT Connectivity*. At that IoT station, we covered one important technical aspect of IoT solutions, which is *IoT connectivity*. We covered the different types of IoT connectivity options that you could use in your large-scale IoT solution.

Chapter 2 Quick Message

We covered so many things in *Chapter 2, The "I" in IoT – IoT Connectivity*, but if there is one important and quick message we would like to share with you from that chapter of the many messages we highlighted, it would be that you, as an IoT solution architect, should study very well where IoT devices and IoT Edge will be located and how they will communicate with each other and with the IoT backend cloud solutions. You should also study the factors that affect the selection of IoT connectivity options, such as device power capability (low-power devices or not?) and device communication modules (what kind of connectivity the IoT devices and IoT Edge devices support). Accordingly, and with the other details discussed in *Chapter 2, The "I" in IoT – IoT Connectivity*, you will be able to select the best IoT connectivity options for your IoT solution.

Then the IoT train moved on to the next IoT station which was *Chapter 3, The "T" in IoT – Devices and Edge*. At that IoT station, we covered another important technical aspect of IoT solutions, which was IoT devices or things. We covered at a high level the different technical aspects of IoT devices and IoT Edge from a hardware and software perspective. We did not go into electronics details, such as microcontroller design or **Printed Circuit Boards** (**PCBs**); however, we covered the selection criteria that you, as an IoT solution architect, can use to evaluate and select the best IoT device/microcontroller for your IoT solution.

Chapter 3 Quick Message

We covered so many things in *Chapter 3, The "T" in IoT – Devices and Edge*, but if there is one important and quick message we would like to share with you from the chapter of the many messages we highlighted, it would be that IoT devices or IoT things are the core of any IoT solution (remember, this is the letter T in IoT), hence, selecting the right IoT devices is a day-zero task for any IoT solution architect or designer. So you must select IoT devices that are secure (by design), resilient, and rich in terms of features they support, such as different connectivity modules, power management, and different security features. Also, it is not only the hardware part that you should take care of but also the software or firmware of such devices. Finally, edge computing is the future, hence you should design the IoT solution with IoT Edge in mind and leverage what IoT Edge can provide and solve in different business domains.

Then the IoT train moved on to the next IoT station, which was *Chapter 4, Diving Deep into the IoT Backend (the IoT Cloud)*. In that chapter, we looked at the IoT backend or IoT Cloud, covering mainly the infrastructure part of the IoT Cloud, and explored the different infrastructure options that you can select for the IoT Cloud of your IoT solution.

> **Chapter 4 Quick Message**
>
> We covered so many things in *Chapter 4, Diving Deep into the IoT Backend (the IoT Cloud)*, but if there is one important and quick message we would like to share with you from that chapter of the many messages we highlighted, it would be that the IoT backend or IoT Cloud is the most important component in your overall E2E IoT solution. The more this component is robust, resilient, scalable, and secure, the more your overall IoT solution will be resilient, scalable, performant, highly available, and secure, so you must select the best hardware, infrastructure, and hosting services for your IoT backend cloud.

Then the IoT train moved on to the next IoT station, which was *Chapter 5, Exploring IoT Platforms*. In that chapter, we continued our sub-journey about the IoT backend or IoT Cloud, but this time we covered the software part of the IoT backend cloud and explored the different software solution components of the IoT Cloud that are used as a backbone for large-scale and production-grade IoT solutions.

> **Chapter 5 Quick Message**
>
> We covered so many things in *Chapter 5, Exploring IoT Platforms*, but if there is one important and quick message we would like to share with you from that chapter of the many messages we highlighted, it would be that the IoT backend or IoT Cloud contains lots of software solutions and components. We first explained those components in a generic way, just in case you decide to build IoT Cloud components and solutions from scratch in a private cloud or public cloud. However, we recommended a managed services approach, hence we proposed using a managed IoT Cloud solution or platform.

Then the IoT train moved on to the next IoT station, which was *Chapter 6, Understanding IoT Device Management*. In that chapter, we covered an important component and it is rare to find a large-scale or production-grade IoT solution that does not use it – IoT device management, which we looked at in some detail.

> **Chapter 6 Quick Message**
>
> We covered so many things in *Chapter 6, Understanding IoT Device Management*, but if there is one important message we would like to share with you from that chapter of the many we highlighted, it would be that "use or buy", instead of building an IoT device management solution that is scalable, has a rich set of device management features, supports different device management standard protocols such as LwM2M, and is flexible and extendable.

Then the IoT train moved on to the next IoT station, which was *Chapter 7, In the End, It Is All about Data, Isn't It?*. In that chapter, we looked at one of the greatest goals of any IoT solution, which is gaining business insights and performing recommended actions based on the IoT data collected from the physical world. We explained IoT data analytics solutions and how you can design and build one.

> **Chapter 7 Quick Message**
>
> We covered so many things in *Chapter 7, In The End, It Is All about Data, Isn't It?*, but if there is one important message we would like to share with you of the many messages that we highlighted, it would be that data is usually the ultimate business goal of any IoT solution, so you must first know what kind of data you need to collect through IoT technologies, and then what you will do with such data to gain the desired business benefits and outcomes.

Then the IoT train moved on to the next IoT station, which was *Chapter 8, IoT Application Architecture Paradigms*. In that chapter, we covered the different modern software app architecture paradigms and technologies that are typically used in building robust, performant, highly available, resilient, and secure IoT backends that serve or act as the backbone for whole IoT solutions.

> **Chapter 8 Quick Message**
>
> We covered so many things in *Chapter 8, IoT Application Architecture Paradigms*, but if there is one important message we would like to share with you of the many messages that we highlighted, it would be yes, there are lots of hyperscale and production-grade IoT platforms that provide managed or serverless IoT backend services that you can use for your IoT solution, but in some cases, you will find yourself as an IoT solution architect having to design and deliver either additional components to ready-made IoT platforms or even build a whole IoT backend/platform from scratch. This is why modern app and cloud-native architectures and solutions are such important topics for any IoT solution architect.

Then the IoT train moved on to the next IoT stop, which is *Chapter 9, Operational Excellence Pillars for Production-Grade IoT Solutions*. On that IoT stop, we covered the different IoT operational excellence pillars, such as high availability, resiliency, monitoring, security, and automation of production-grade and large-scale IoT solutions.

Chapter 9 Quick Message

We covered so many things in *Chapter 9, Operational Excellence Pillars for Production-Grade IoT Solutions*, but if there is one important message we would like to share with you of the many messages that we highlighted, it would be that the challenge is not only in designing and building large-scale IoT solutions; the challenge is also in operating such large-scale and production-grade IoT solutions, so aspects such as high availability, resiliency, security, and performance must be considered from day zero. In other words, they should be considered during the design phase of such large-scale and production-grade IoT solutions.

And finally, the IoT train came to this chapter, *Chapter 10, Wrapping Up and Final Thoughts*, which highlights the end of that IoT journey in this book.

In this chapter, we will cover the following topics:

- Summary using one IoT reference architecture
- IoT and future trends

Summary using one IoT reference architecture

The best way to summarize what we have explained in this book's chapters is to go through one E2E IoT solution architecture and highlight where you can find more information about each component or layer of that E2E architecture in the chapters.

As stated throughout the book, the concepts we discussed are generic, so if you are using the Microsoft Azure Cloud and its IoT platform, Google Cloud and its IoT platform, or the AWS Cloud and its IoT platform, you will find lots of similarities (that is, the IoT Edge cloud, the IoT Core cloud, managed services, and solutions from the cloud that are used in building different software components and solutions (data analytics, ML, AI, apps, and so on) on their proposed and recommended IoT reference architecture).

For example, here is one **Azure IoT reference architecture**:

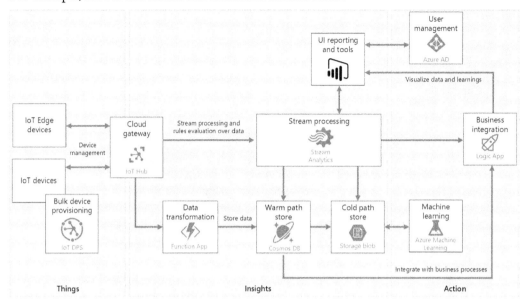

Figure 10.1 – Microsoft Azure IoT reference architecture

And here is the **Google Cloud IoT reference architecture**:

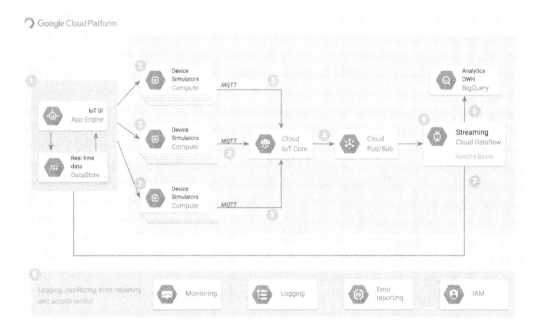

Figure 10.2 – Google Cloud IoT reference architecture

But since we used **AWS** and its **AWS IoT platform** throughout this book as a reference in our IoT journey, we will use an example of AWS IoT reference architecture to explain or to summarize, as far as we can, what has been explained in the previous chapters.

We have selected an IoT reference architecture that is used in industrial IoT. The top IoT use cases for the industrial sector are typically related to process improvement and automation, predictive maintenance, and asset tracking and monitoring. The reference architecture that we will use in this chapter – as shown in *Figure 10.3* – is related to an anomaly detection use case, which can be classified under the predictive maintenance use case:

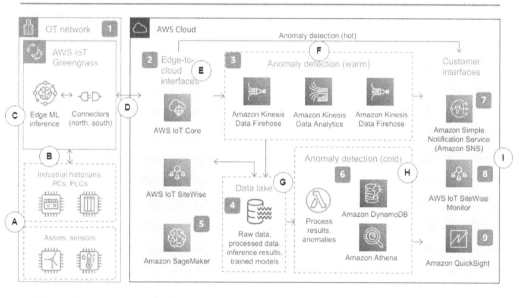

Figure 10.3 – AWS IoT reference architecture

Let's walk through this architecture (note: the letters in the circles in *Figure 10.3* match the points that follow):

A. In any industrial site or factory, you will find lots of IoT devices or nodes (aka sensor nodes), sensors (for example, temperature, power, flow, pressure, level, and proximity), actuators (for example, motors, relays, and solenoids), industrial historians (that is, data stores for the data generated in the production plant/factory during the manufacturing processes, which are typically massive sets of data – thousands of bits of data are generated each second), **Process Control Systems (PCSs)**, **Programmable Logic Controllers (PLCs)**, and other devices that are deployed in the factory site to collect data and control the manufacturing processes.

In this book, we explained, in *Chapter 3, The "T" in IoT – Devices and Edge*, IoT devices and the IoT Edge at a high level. We explained microcontrollers and the different aspects of what makes up an IoT device or a sensing node in terms of hardware, software, and communication modules.

B. Typically, in an industrial site, the sensing nodes are controlled by a controller (PLCs). Now, to communicate with such controllers, the most common, standard, and widely used communication protocol is **Open Platform Communications Unified Architecture (OPC UA).**

In other IoT scenarios (that is, consumer IoT) such as smart homes, you might have sensing nodes or IoT devices using different protocols for communication, such as Zigbee, Z-Wave, and Bluetooth. But since we are talking about an industrial scenario here, you will typically find OPC UA is used (there are some other industrial communication protocols, but they are either proprietary or older protocols than OPC UA).

In *Chapter 2, The "I" in IoT – IoT Connectivity*, we covered IoT connectivity and the different connectivity options that are typically used in large-scale IoT solutions.

The controllers need to send the data to the IoT gateway/IoT Edge, which is typically deployed on the same site or very close by – see point C.

C. The AWS implementation of IoT Edge software, as explained in *Chapter 3, The "T" in IoT – Devices and Edge*, is the **AWS Greengrass service**, which has many features. But in *Figure 10.3*, two features are used in that architecture, **connectors** and **ML inference** at the edge.

The connectors are for northbound and southbound integration, meaning the southbound connectors in this architecture are the AWS SiteWise connectors that collect data from the industrial historian, PLCs, and so on using a protocol such as OPC UA, while the northbound connectors are the connectors that are integrated with AWS Cloud services such as the AWS SiteWise cloud service and AWS IoT Core (see edge-to-cloud interfaces in the diagram – rectangle number 2).

The IoT gateway sits in between the industrial site or the **Operational Technology (OT) network** (rectangle number 1 in the diagram) and the AWS Cloud, which is accessible over the internet.

The northbound connector's job is to stream the IoT data collected in the site or factory to the IoT Cloud (the AWS Cloud – see point E). In other words, it is the connector that interfaces with AWS Cloud services.

The other feature used in that architecture in **AWS Greengrass** is the **ML inference** feature.

We explained in *Chapter 3, The "T" in IoT – Devices and Edge*, that you typically *train* and *build* an ML model in the cloud, where you have the massive compute infrastructure and storage that are required to do that task. Once you have the ML model ready, then you can deploy it at the edge so you can have real-time inference. In other words, you can call the model APIs at the edge and get an immediate response.

Without going into details and to simplify this machine learning part, the idea of ML at a high level is the following: in the traditional software programming paradigm, you know the model or the algorithm beforehand, for example, you have a software program that multiplies any input it receives by 2, adds 1 to the result of that multiplication, and then outputs the result. So let's assume the input variable is X and the output variable is Y. Then, the model, equation, or algorithm will be Y=(X*2)+1. So, let's test/train that model. If the input (X) to the program was 2, then the output would be (2*2)+1, so it would be 5, which is true. If X was 3, the output (Y) would be 7, and so on.

In the machine learning paradigm, it is the opposite process, meaning we have the data of the output or outcome of some unknown process/algorithm. That output data was generated based on some parameters or inputs. The machine learning's job is to figure out the unknown model or algorithm that makes that relationship between the output or outcome data and those parameters or inputs. In the example we mentioned earlier, Y=(X*2)+1, let's assume we don't know that equation or model, but we know the value of X that generates the value of Y. In other words, we don't know the equation of X has to be multiplied by 2 and then 1 added to that multiplication to get the value of Y.

So, if we take a 10-year-old child and give them a series of numbers such as 1, 3, 5, 7, A, B, ..., and N, and we ask them to figure out what could be the values of A and B in that series, they will try first to figure out the pattern (or the model) of that series and then, once the model/pattern is found, they can infer or try the discovered model/pattern to figure out the values of A and B (and any other value in that series).

By looking into this series of numbers, they might figure out the model/pattern as with any number starting from the zero position. If I multiply that position by 2 and then add 1 to it, it will give me the value of that number position in this series.

Great – we have the model/pattern/algorithm now (outcome = (position of the number * 2) + 1). To make sure that the model is correct, you need to train it by checking the values of some given numbers in that series, for example, at position 0, it will be (0*2)+1, and then the output value will be 1. Excellent – this is true (this is the first item in the series). You continue with a position equal to 1 (it will be (1*2)+1, and then the value will be 3). Great – this is also true as the value at position 2 is 3 in the series and so on. By training that model, you figure out the accuracy of the model (that is, how many times you get the output value true). Here, the accuracy of that model is 100% because we picked a very basic and simple case for the sake of explanation.

Now, once you have the model and you have trained it, you can answer the question about the values of A and B. For A, it will be 9 (since its position is 4), for B, it will be 11 (since its position is 5), and so on. Getting the value of A and B is known – in the ML world – as **ML inference**.

So, in short, and back to machine learning, in the end, the ML model will be a software program that holds the equation or formula that is created based on the training of massive data that was generated earlier based on some parameters and inputs. The ML program or model can be hosted in Edge, but to create and train that model, you need a massive compute and data storage infrastructure, hence the building and training part of the ML model is typically done in the cloud, not in Edge (note, the example we mentioned is just for simplicity; otherwise, industrial and production ML models are typically very complex and require lots of training, evaluation, and optimization).

Again, without going into the details of ML models, the model to detect anomalies can be understood – at a high level – by the following example: If you know beforehand a rule such as whether the temperature reported by the IoT device is greater than the normal value, then you know you can classify that as an anomaly. I used the temperature parameter here for simplicity. Other parameters in other business domains can be used to identify anomalies. This is one easy way to discover anomalies. You don't need an ML model to detect such anomalies.

But if you don't have a rule beforehand to identify anomalies, then machine learning models can help in identifying them by leveraging one of the clustering algorithms used in ML. For example, without going into details, think about thousands or millions of IoT devices. The majority of those devices have a similar pattern in terms of internet data usage (that is, the amount of internet data consumed daily, its frequency, and so on), but a few of those devices or even one device has something completely or partially different. For example, the typical or normal internet data consumption by normal IoT devices is a few bytes or megabytes per day, but if one or a few IoT devices consume gigabytes of data per day, then clearly there is something wrong with that device, cluster, or group of IoT devices and someone needs to check what is going on. It might be the case that those IoT devices have been stolen from their location and someone is using their internet connectivity to do internet surfing, for example, watching YouTube or visiting social media websites, or perhaps even that the IoT device has been hacked and participated in a DDoS attack and so on. An ML anomaly detection model will help you in identifying such anomalies by leveraging one of the ML clustering algorithms.

There are so many ML frameworks that data scientists use to build, train, and evaluate different ML models. AWS provides – on the AWS Cloud, not Edge – a fully managed ML service called **Amazon SageMaker** (see point E in the diagram and in this list). This service helps data scientists in building, training, and evaluating their ML models.

The good news here is that with AWS, Azure, or Google Cloud, you will find a huge and rich set of ready-made ML and AI services that you can just use directly without building, training, or evaluating your own ML model. Anomaly detection, for example, is one of those ready-made ML models that you can use directly in your IoT solution.

To conclude this point, an ML model is built, trained, and evaluated in the AWS Cloud and can be deployed and inferenced/called at Edge.

D. Before we jump into the AWS Cloud side, let's talk about the **connectivity** between the edge cloud (or the OT network or factory side) and the AWS Cloud that hosts different cloud services, such as AWS IoT Core, AWS IoT SiteWise, and AWS SageMaker. Usually, such connectivity is done over the internet. You can have a secure connection though by building a VPN/IPSec (over an internet connection) between the edge network and the IoT Cloud network, or have a private dedicated connection (lease line) between the edge network and the IoT Cloud network (in AWS, services such as AWS Direct Connect can be used for that private connection purpose).

In *Chapter 2, The "I" in IoT – IoT Connectivity*, we explained the purpose of IoT gateway devices and how they support different connectivity options. One connectivity option or communication module is installed in the IoT gateway device to handle the connectivity between the IoT gateway device from one end and the IoT endpoint/leaf devices on the Edge/factory site from the other end (for example, the IoT device gateway has a Zigbee communication module to communicate with IoT endpoint devices that also have a Zigbee communication module). The other connectivity option or communication module that is typically also installed into the IoT gateway device is used to handle the communication between the IoT device gateway from one end and the IoT backend cloud from the other end over the internet connection or over a dedicated private connection (for example, the IoT device gateway has an LTE/cellular communication module and that module gives the IoT device gateway access to the internet so the IoT device gateway can relay/forward the messages it received from the IoT endpoint devices through the Zigbee interface to the IoT backend cloud over the internet).

E. AWS perfectly calls that layer (rectangle number 2 in the diagram) in this reference architecture Edge-to-cloud interfaces as those three services, AWS IoT Core, AWS IoT SiteWise, and Amazon SageMaker, are the AWS Cloud-based services that interact with edge services such as AWS IoT Greengrass through different AWS IoT Greengrass connectors.

We covered **AWS IoT Core** in *Chapter 5, Exploring IoT Platforms*, and we covered **AWS IoT SiteWise** in *Chapter 7, In the End, It Is All about Data, Isn't It?*. **Amazon SageMaker** is beyond the scope of this book.

AWS IoT Core will get the streamed IoT data of the factory site from the edge (Greengrass) and then, in the AWS Cloud, through a component such as **AWS IoT Rules Engine** (not shown in the reference architecture diagram), such IoT data will be sent to two different data paths, a warm data path (rectangle number 3 or point F on the diagram) and a cold data path (rectangle number 4 or point G on the diagram). We explained that concept (forking streaming data into hot and cold storage) in *Chapter 7, In the End, It Is All about Data, Isn't It?*, when we talked about the **Lambda architecture** that split the batch and streaming processing in the IoT data analytics domain.

F. **Hot data path layer**: In this layer, we have the following components:

- **Amazon Kinesis Data Firehose**: This component has two jobs to do in the architecture. First, it receives the IoT data that is coming from IoT Core through AWS IoT Rules Engine, as explained in the previous point. It buffers such IoT data and can also do some data enrichment and transformation – if needed – on that IoT data by calling a Lambda function that can do that data enrichment or transformation. Then, Amazon Kinesis Data Firehose sends such buffered (and enriched) IoT data to the **Amazon Kinesis Data Analytics** service for further processing (applying near real-time data analytics). The other job of Amazon Kinesis Data Firehose is to send the outcome of the data analytics job that is done by the Amazon Kinesis Data Analytics service to a destination such as Amazon S3 (if you notice in the diagram, the data lake contains raw data, processed data, and so on. This processed data is, in fact, what was generated from the Amazon Kinesis Data Analytics near-real-time analytic application, and the outcome from that near-real-time data analytics application is sent to Amazon S3 through Kinesis Data Firehose). Amazon Kinesis Data Firehose can send data records to various destinations, including Amazon Simple Storage Service (Amazon S3), Amazon Redshift, the Amazon OpenSearch service, and any HTTP endpoint.

- **Amazon Kinesis Data Analytics**: This is the data analytics engine that does the near-real-time data analytics. In that component, you build a near-real-time SQL application by writing SQL statements that process input (from Kinesis Data Firehose) and produce output (to Kinesis Data Firehose as well). In the Amazon Kinesis Data Analytics application, you can write SQL statements against in-application streams (Amazon Kinesis Data Firehose) and reference tables. You can also write join queries to combine data from both sources. The outcome of that layer is either processed data that is stored in a data lake (Amazon S3) for further analysis (to be used for cold anomaly detection) or the triggering of a near -real-time notification (email, SMS, and so on) for end users using Amazon Simple Notification Service (Amazon SNS) – see the customer interfaces layer or point I in the diagram. To conclude the work that is done in this layer, the IoT data of the factory site is buffered for a short time window (5, 10 minutes, and so on) in Amazon Kinesis Data Firehose, and then SQL queries are run (as part of the Amazon Kinesis Data Analytics application) against that small window of data to do near-real-time analytics. The output of those near-real-time analytics is either stored in the data lake for further analysis (batch processing, historical analytics, and so on) or a near-real-time notification (email, SMS, and so on) is sent to end users using the Amazon SNS service. In *Chapter 7, In the End, It Is All about Data, Isn't It?*, we discussed IoT data analytics in detail.

G. **The data lake**: This is where all IoT data is stored –raw data (data coming from IoT devices on the site), processed data (data coming from Amazon Kinesis Data Analytics or Amazon Kinesis Data Firehose following some data enrichment), and ML inference results (data coming from AWS IoT Greengrass through its connector of ML inference that happened on the edge). In other words, the result of ML inference that occurred on the edge is sent to the data lake for further processing (to re-train and optimize the accuracy of the anomaly detection model). To optimize the model, you need to know the results you got from the deployed ML model and check the inference result accuracy of that model to see how you will enhance/optimize the model accuracy further). And finally, the ML-trained models are also stored in Amazon S3 (in a data lake).

H. **Cold data path layer**: This is the layer where we do the batch processing on the raw data, processed data, and ML inference result data to detect any anomalies in such big data. Tools such as Amazon Athena (an analytic query engine that runs queries on top of S3 stored data) and Amazon Lambda functions can be used in executing that batch processing.

I. **Customer interface layer**: This contains the services that interact with the customers or end users, such as the AWS SNS service, which is used for sending notifications to customers or operations teams; AWS IoT SiteWise Monitor, a ready-made web application that operations teams can use to monitor industrial site assets – refer to *Chapter 7, In the End, It Is All about Data, Isn't It?*, when we talked about AWS IoT SiteWise; and finally, Amazon QuickSight, a service used for dashboards, reports, and visualizations.

This was a quick explanation of an industrial IoT solution architecture example to remind you of and summarize the different topics that we have covered in this book.

In the next section, let's discuss the future of IoT and the trends and technologies that you, as an IoT solution architect, should be aware of and keep an eye on.

IoT and future trends

IoT is already hyped now and it will keep booming and growing in the future with advancements in IoT ecosystem technologies.

As we stated in the first chapter of this book, *Introduction to IoT – The Big Picture*, IoT technologies are key enablers for digital transformation and they are also enablers for other technologies such as AI, ML, and advanced data analytics, but the question here is whether any technologies will impact or shape the future of IoT technologies? The answer is yes. Let's explain this in more detail next:

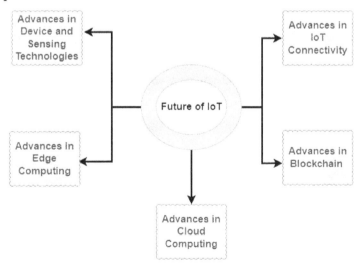

Figure 10.4 – Technologies that impact the future of IoT

From our point of view, advances in the following technologies are what will shape the future of IoT:

- **Devices and sensing technologies**: Advances in microcontrollers (in terms of size, capacity, power management, peripherals, and so on) and sensors (in terms of sensitivity, resolution, linearity, zero drift and full-scale drift, range, repeatability, and so on) will clearly have a big impact on the IoT domain as that advancement will lead to an increase in the number of IoT-connected devices, which, in the end, will give businesses and consumers more visibility and insights into the physical world around us.

  IoT connectivity: Advances in IoT wireless connectivity in terms of speed in data transmissions, lower latency, high capacity, and throughput will also lead to an increase in the number of connected devices and an increase in interesting consumer and industrial apps and solutions such as augmented reality (VR/AR), remote surgery, digital twins, and so many other interesting apps and solutions in different business domains that we have mentioned earlier, in *Chapter 1, Introduction to The IoT – The Big Picture*.

  In IoT cellular connectivity, everyone nowadays is talking about **5G** and the benefits that it brings in terms of data speed, latency, and capacity for IoT industrial and consumer applications.

  5G (and future 6G) connectivity options are shaping the future of **high-power IoT devices** used in industrial and consumer IoT use cases (for example, security video surveillance with **Ultra High-Definition** (**UHD**) video such as 4K and 8K that can be transmitted with minimal delays, smart factories with high-definition video of machines, and different factory assets that are transmitted instantly to the cloud for inspection or maintenance, remote surgery, and so on).

  Another advanced feature that was kind of born with 5G and will have an impact on the future of IoT is **network slicing**, which, in short, is a virtualized telecommunication network. In other words, a telecommunication provider can create a dedicated network for enterprises where those enterprises will have their own dedicated telecommunication network (starting from radio access, the core network, and business and operational support systems (BSSs, OSSs).

The **Radio Access Network (RAN)**, **core network**, and **BSS/OSS** were (and still are) commonly shared resources between enterprises/consumers. Now, with network slicing and other technologies, you can have a slice of public radio/core network/BSS dedicated to a specific enterprise or organization. For example, in a factory, you can have a complete and dedicated telecommunication network that covers and is dedicated to that factory only.

Network slicing has lots of benefits, such as the efficient and effective utilization of network resources as network operators can allocate the right amount of required resources as per the network slice. For example, one network slice can be designed to deliver low latency and a low data rate, while another network slice can be configured to deliver high throughput and so on. This is a great benefit as we know that IoT use cases are usually different, so based on your IoT use case, you can ask for a dedicated network slice with a specific network characteristic that fits with your case.

In other words, with network slicing, mobile operators or telecommunication providers can now provide customizable network capabilities with custom (based on customers' needs) data speed, quality, latency, reliability, security, and services.

Advances in **Low-Power Wide-Area (LPWA)** technologies, **cellular LPWA** (NB-IoT and LTE-M), and non-cellular **LPWA** (LoRa and LoRaWAN) for low-power IoT devices will also keep evolving and growing in the future to serve many **low-power IoT use cases** in industrial and consumer segments (for example, smart metering, smart parking, grid management, smart building, smart agriculture, asset tracking, consumer-wearable gadgets, fleet management, and usage-based insurance).

- **Cloud computing**: The more IoT devices are connected, the more data will be generated and collected, and the more compute you will need to process such big data. Therefore, advances in cloud computing in terms of computing and storage capabilities to handle such a massive IoT scale (billions of IoT -connected devices and trillions of bits of IoT data and messages) will have a great impact on the future of IoT.

Not only will advances in computing and storage services have a great impact on the future of IoT, but also the advances in PaaS and SaaS services (for example, IoT platform services, AI, ML, advanced data analytics, robotics, modern app development, and DevOps and automation tools) will play critical roles in the future of IoT in terms of IoT products and solution agility, faster time to market, and operational excellence.

- **Edge computing**: The advancement in edge computing in terms of edge location (becoming closer to IoT devices or end users), edge device hardware capabilities (the size of edge devices becoming smaller and yet more powerful in terms of compute and storage to help in deploying it anywhere), and edge software (more software services to run in Edge) will have a great impact on the future of IoT.

  As of now, we see great edge computing services and features that are already used and applied in many IoT solutions – services such as machine learning, AI, and near-real-time data analytics services. In the future, we expect additional services to be added, hosted, and run on the edge cloud that will eventually enrich and advance the different IoT solutions and services that are delivered to enterprises and consumers.

- **Blockchain**: Having a shared, distributed, and immutable ledger (a special type of data store) that can only be accessed by members of that blockchain network with permission is the core feature or the heart of the blockchain technology. In other words, in the blockchain network, a record or a transaction, once it is created in the blockchain, can't be altered. That record is stored identically in multiple locations and has time and date stamps.

  Because of such characteristics, blockchain technology introduced the following benefits:

  i. **Enhanced security**: Hacking systems to expose their data became very difficult with blockchain technology as data is replicated and stored in different locations and is immutable, meaning any kind of fraud/tampering will be easily detected and eliminated.

  ii. **Enhanced traceability and auditing**: Blockchain creates an audit trail of the transactions entered in the blockchain network, which increases traceability and auditing.

  iii. **Enhanced automation and speed**: Blockchain technology removes the barriers introduced by manual work and paperwork, which are time-consuming and prone to human error. With Blockchain – with a smart contract use case, for example – there is no need for a third party to verify transactions, reconcile multiple ledgers, exchange papers, and so on.

Blockchain and IoT

Blockchain technologies and IoT technologies have both separately disrupted and added business value and benefits to many industries, such as supply chains, the food chain, the financial industry, healthcare, the pharmaceutical industry, insurance, and the government sector. When both technologies combine or work together, more business benefits and values will be added. IoT technologies, on the one hand, are capable of bringing "data" from the physical world, while blockchain technologies, on the other hand, are capable of securing such collected IoT data, and enhance the user and customer experience by removing lots of barriers introduced by manual work and paperwork.

To give an example of how Blockchain works with IoT, in the supply chain sector, for example, the track-and-trace feature is very challenging. Products are usually shipped from the factory in excellent condition, but then what happens to those products between them leaving the factory and reaching consumers is unknown to business owners. The business might lose out in that case because products might reach the consumers broken or in a bad condition. This is a business problem that needs a solution.

For that business problem, IoT technologies can help by having sensors deployed in shipment containers to catch and collect the container's temperature, humidity, location, and so on. But how can business owners trust, or make sure, that the data collected is trustworthy and does not change on a long journey (from the factory to the consumer), especially as supply chain processes are typically so long and paper-based? The business still has no control or visibility over what might happen with such manual/human work on that long journey. For this challenge, blockchain technologies can help by removing those manual work and paperwork processes and ensuring the trustworthiness of the IoT data collected from the moment the products leave the factory until they reach consumers.

Another example of what IoT and Blockchain could do together is a new trend called **Economy of Things (EoT)**. That trend is driven mainly by IoT and blockchain technologies.

In **EoT**, IoT devices play a role in the digital market/economy. How? Remember the smart parking example use case that we covered in *Chapter 1, Introduction to The IoT – The Big Picture*. We covered many use cases for the smart parking solution, but if you remember, there was one use case for the driver (human) to reserve and pay for a free parking lot that IoT helped in finding, right? In EoT, IoT devices will take it even further by reserving and paying for that parking lot on behalf of the driver. You might be wondering how such transactions – between an IoT device (a connected car, in that example) and a smart parking payment gateway – would be done securely. The answer is that they would be done with the help of blockchain technology due to the characteristics we mentioned earlier in relation to Blockchain.

EoT is another stream or source for **IoT monetization**. Monetizing IoT data still represents a big chunk of the IoT monetization strategy.

In the healthcare sector, Blockchain helps, and will help, IoT technologies a lot with securing, storing, and managing patient data collected from the IoT devices attached to patients or medical equipment.

In the insurance and food chain sectors and so on, Blockchain helps and supports different IoT solutions by increasing transactions' trust, security, transparency, efficiency, and speed, as well as by saving costs.

Those are the key technology trends that you, as an IoT solution architect or designer, should be aware of. Keep watching the advancement that happens in those technologies to reflect that in your large-scale, future-proof, and production-grade IoT solutions.

With that, we have come to the end of this IoT journey in this book. We hope you have enjoyed it and learned many things in the different domains and technologies of IoT. Please let us know how we could improve this book in the future by reaching out to us via the methods mentioned in the *Preface* section.

Index

B

D

`Packt.com`

Subscribe to our online digital library for full access to over 7,000 books and videos, as well as industry leading tools to help you plan your personal development and advance your career. For more information, please visit our website.

Why subscribe?

- Spend less time learning and more time coding with practical eBooks and Videos from over 4,000 industry professionals

- Improve your learning with Skill Plans built especially for you

- Get a free eBook or video every month

- Fully searchable for easy access to vital information

- Copy and paste, print, and bookmark content

Did you know that Packt offers eBook versions of every book published, with PDF and ePub files available? You can upgrade to the eBook version at `packt.com` and as a print book customer, you are entitled to a discount on the eBook copy. Get in touch with us at `customercare@packtpub.com` for more details.

At `www.packt.com`, you can also read a collection of free technical articles, sign up for a range of free newsletters, and receive exclusive discounts and offers on Packt books and eBooks.

Other Books You May Enjoy

If you enjoyed this book, you may be interested in these other books by Packt:

Developing IoT Projects with ESP32

Vedat Ozan Oner

ISBN: 9781838641160

- Explore advanced use cases like UART communication, sound and camera features, low-energy scenarios, and scheduling with an RTOS

- Add different types of displays in your projects where immediate output to users is required

- Connect to Wi-Fi and Bluetooth for local network communication

- Connect cloud platforms through different IoT messaging protocols

- Integrate ESP32 with third-party services such as voice assistants and IFTTT

- Discover best practices for implementing IoT security features in a production-grade solution

Intelligent Workloads at the Edge

Indraneel Mitra, Ryan Burke

ISBN: 9781801811781

- Build an end-to-end IoT solution from the edge to the cloud
- Design and deploy multi-faceted intelligent solutions on the edge
- Process data at the edge through analytics and ML
- Package and optimize models for the edge using Amazon SageMaker
- Implement MLOps and DevOps for operating an edge-based solution
- Onboard and manage fleets of edge devices at scale
- Review edge-based workloads against industry best practices

Packt is searching for authors like you

If you're interested in becoming an author for Packt, please visit `authors.packtpub.com` and apply today. We have worked with thousands of developers and tech professionals, just like you, to help them share their insight with the global tech community. You can make a general application, apply for a specific hot topic that we are recruiting an author for, or submit your own idea.

Share Your Thoughts

Now you've finished *Designing Production-Grade and Large-Scale IoT Solutions*, we'd love to hear your thoughts! Scan the QR code below to go straight to the Amazon review page for this book and share your feedback or leave a review on the site that you purchased it from.

https://packt.link/r/1838829253

Your review is important to us and the tech community and will help us make sure we're delivering excellent quality content.

www.ingramcontent.com/pod-product-compliance
Lightning Source LLC
Chambersburg PA
CBHW081504050326
40690CB00015B/2921